T0244120

THE
BEST
OF
ALL
POSSIBLE
WORLDS

THE
BEST
OF
ALL
POSSIBLE
WORLDS

A Life of Leibniz
IN SEVEN PIVOTAL DAYS

MICHAEL KEMPE
TRANSLATED BY MARSHALL YARBROUGH

W. W. NORTON & COMPANY

Independent Publishers Since 1923

First published in Germany in 2022 as *Die beste aller möglichen Welten*
by S. Fischer Verlag in Frankfurt am Main.

German text © 2022 by S. Fischer Verlag GmbH,
Hedderichstr. 114, D-60596 Frankfurt am Main, Germany
English translation © 2024 by Marshall Yarbrough

The translation of this book was supported by a grant from the Goethe-Institut.

All rights reserved
Printed in the United States of America
First Edition

For information about permission to reproduce selections from this
book, write to Permissions, W. W. Norton & Company, Inc.,
500 Fifth Avenue, New York, NY 10110

For information about special discounts for bulk purchases, please
contact W. W. Norton Special Sales at specialsales@wwnorton.com or
800-233-4830

Manufacturing by Lakeside Book Company
Book design by Lovedog Studio
Production manager: Anna Oler

ISBN 978-1-324-09394-7

W. W. Norton & Company, Inc., 500 Fifth Avenue, New York, NY 10110
www.wwnorton.com

W. W. Norton & Company Ltd., 15 Carlisle Street, London W1D 3BS

1 0 9 8 7 6 5 4 3 2 1

FOR
Clara Emilia,
Christiane,
AND
a little fly

CONTENTS

Machines—Autobiographical Flashbacks—Blurred Infinities
—Soul-Deserts—How Many Leibnizes in One Person?—
Life-Forms in Flux—Monads 2.0

In Search of a Global Formula—Summer of Love—On
Brüderstraße—Live Transmission—Report from Europe—
Learning from China—Digital Dreams—A World of Zeroes
and Ones—A Chinese Oracle in Binary Structure—Magic
Square with Dyadic Symbols—Structures of Reality—
Transfer into the Future

All a Matter of Perspective—Outside the Library Door—
Fur Stockings and Felt Socks—The Historian's Tools—
A Woman on the Papal Throne?—God on Trial—Finding
the Positive in the Bad—Wroth and Brutal, the Last in the
Line of the Roman Kings—Two Figures Caught between
Truth and Fiction—The Progression of Time

Deep Breath . . . and Another—Success and Solitude—
The Distant Proximity to Power—Strained Long-Distance
Relationships—Turbulent Days—Suitcases Packed—Love
and Geometry—A Calculator to Rival God—Souls without
Windows?—Tomorrow Never Knows

Chapter 7: HANOVER, JULY 2, 1716
Running Headlong into the Future:
Spiral-Shaped Progress and Post-human Intelligence...... *179*

Thinking beyond One's Self—Bursting with Life—To the
Springs—For a New Europe—A Good Mood, Spoiled—
Polymaths, Kindred Spirits—How Old Is the Earth?—The
Salamander Who Came In from the Heat—Can There
Have Been a Beginning?—The Return of Yesterday—
Understanding the Flies—A Day in the Life

INTRODUCTION

WHY IS THERE A WORLD AT ALL, AND WHY THIS ONE? IT'S not a bad place to start things off. The question catapults us right into the middle of the labyrinthine thought of Gottfried Wilhelm Leibniz. Among the infinite number of possible worlds, there is one that is best, runs Leibniz's answer—otherwise God would never have decided to create a world in the first place. And if there were a world better than this one, then God would have created *it* instead of the one we live in. In reading these words, we quickly find ourselves in a world of necessary logical conclusions and rigorous rationality, a world that seems to subject not just people but also God himself to the iron laws of truth and necessity. Today it has become alien to us to speak of the world in such a way. For one thing, ever since Kant made his critique of reason, such statements have come to be considered unreliable instances of the human *ratio* overstepping its bounds. For another, however, it has been a long time indeed since any single person could boast a command of all the sciences and humanities and achieve excellence in several branches of them at once. Thus it is with all the more fascination that we look back at a time when such a thing still seemed possible, and at those individuals who dared make the attempt. One of them was Leibniz, a polymath or universal genius—as many would consider him—and perhaps one of the last of his kind. In any case, he is a representative of a lost time: the early European Enlightenment, with its roots in the Baroque era. It was a time marked by a rational optimism and a belief in progress, by the confident assurance that humanity would

stride forth into a bright future founded on the basis of a rational religion and driven forward by the achievements of science and
technology. Today, given the critical lens through which we view
progress and growth, given secularization and the stark differentiation that now exists among the sciences, it seems impossible to
reconnect with the world of thought and ideas of Leibniz's time.

Why a world, then, and why this one in particular? Today Leibniz's answer to this question is known mainly through Voltaire's
1759 satire *Candide*, which has almost completely overshadowed it.
As a result, it seems possible to speak of the best of all possible
worlds only if one criticizes and dismisses the proposition, or mocks
it and takes it to ridiculous extremes. Whether Voltaire's critique of
Leibniz truly hits home is today a matter of debate. In order for us
to understand Leibniz's world, his view of things, and also the age
that produced such a view, it's necessary to set the critique aside and
take a step back. The prospect of encountering Leibniz in the midst
of his day-to-day life, of observing him in the act of thinking and
creating, doesn't just promise a thrilling journey into a fascinating
time. It also makes it possible for us to gain a deeper understanding
of his philosophy, his mathematics, and his wide-ranging scientific
endeavors. Suddenly Leibniz doesn't seem so alien at all; indeed, in
many respects he may seem strangely familiar, since people in our
own time are also wrestling with many of the questions and problems that concerned him. From this vantage, the full-bottomed wig
and frock coat no longer seem essential characteristics of an inaccessible figure from a past epoch; they turn out to be mere trappings.
Look past them, and a person steps into view who bears a resem-
blance to the isolated individuals of the present day, constantly communicating and yet simultaneously withdrawn into themselves.

Unhappiness and suffering, people torturing one another, and on
top of all that, horrifying natural disasters like the Lisbon earthquake of 1755: Voltaire has Candide, dismayed, bleeding, and trem-
bling, ask himself, "If this is the best of all possible worlds, what can
the rest be like?"[1] Leibniz views the problem from another angle.

We can no longer go back and change what has been, but we can try to improve what is. Leibniz views the world in terms of its possibilities. Not everything that is possible must or can become reality. But at least some of it can be realized, maybe more than it would sometimes seem. If everything in the world happened out of necessity, then individuals could not be held responsible for their actions; there would be no morality and also no freedom. Not only was God free to choose between different possible worlds; humankind is also, notably, granted the freedom to change the world, to help shape it.

For a long time, this line of thinking has been interpreted solely as abstract reflection of a strict rationalistic bent. Despite frequent claims to the contrary, however, Leibniz was by no means an inflexible rationalist who always had his head in the clouds and dismissed experience. In fact, he had both feet planted firmly in the soil of reality, in his case the reality of the Baroque *Fürstenstaat*, the small state with a prince at its helm. Still, while he did seek the support of the powerful for his ambitious plans to improve the world, at the same time he also sought independence from them so that he could remain free in his scientific pursuits. For Leibniz, freedom has a concrete meaning and a concrete place—or better yet, places. Not just while he is bent over a desk but also as he travels between the courts he serves, a welcome space of freedom opens up for him. For example, when official business takes him to Brunswick and Wolfenbüttel, he takes the opportunity to make a stopover in Ermsleben, a small town south of Halberstadt, and spend a few days visiting Pastor Jakob Friedrich Reimmann. Leibniz enjoys the unstructured hours. He eats at the family table, contents himself with plain cooking, and gives himself over to long learned discussions in Reimmann's study. Leibniz, writes Reimmann looking back, "has on more than one occasion sat up with me till twelve or one o'clock in the morning, chattering away."[2]

Chatting about heaven, earth, and everything in between, deep into the night, as if there were no tomorrow. Every day moves slowly, inexorably toward its end, and a new one begins—not even Leibniz can change that. Accompanying him in his day-to-day life, observing him on seven days in different stages of his life in various places, gives us a chance to better understand his thought and his actions, to more clearly see the connection between his experience of the world and his worldview. On each of these days, we can regard Leibniz from an altered perspective, just as his philosophy calls for constantly changing one's position and taking on a new point of view. Granted, the days selected here can offer only small snippets. Certain things appear only briefly. But what becomes apparent is how Leibniz manages, again and again, to connect far-flung fields of knowledge. Along the way, his dictum that everything is connected to everything else must occasionally double as an excuse for the far-reaching nature of his studies, which often keeps him from finishing anything. A universal scholar simply cannot allow himself to neglect any area of knowledge. One improves the world not only on the grand scale but also with regard to small things, such as by forging a nail shaped in such a way that it becomes wedged in the wood into which it has been hammered—a kind of wall anchor, that is, that Leibniz invents more or less in passing and that today literally holds the world together.[3]

How concrete can or should a philosophy actually get? As we shall see, without coffee or chocolate, at least, Leibniz might not have regarded this as the best of all possible worlds. But it is precisely in the concrete examples that Leibniz's views come to life. Take his contention that God decided to create lions, even though they pose a danger to humans, because without lions the world would be less perfect. Even before Leibniz, variety or diversity was one aspect on which the theologians of Spanish late scholasticism in the seventeenth century, otherwise prone to rather dry and unwieldy arguments, could become astonishingly specific. A world consisting only of flies—so argued the Spanish Jesuit Antonio Perez—is absolutely inconceivable, even if God hadn't been compelled by necessity to create the optimal world.[4] Speaking of the fly, while it doesn't play a large role in Leibniz's philosophy, it does play a small one. For this reason, it also appears occasionally throughout this book: a fictitious fly to begin with, it buzzes off only to suddenly reappear every now and again as an object of Leibniz's thought—like a kind of metaphysical alter-ego for Leibniz himself.

Leibniz shares with the fly his stubborn insouciance. He won't be shooed away; he persistently disturbs the peace of powerful princes to try to win their support for his plans for progress. Conversely, Leibniz never allows the little buzzing pests to keep him from writing. It's hard to fathom that someone who welcomes every distraction, who immediately seizes on every source of stimulation, who postpones nothing, but who tries, rather, to do everything at once, could at the same time be capable of concentrating on reading and writing anywhere and in every situation imaginable. No noise, no stench, no pothole on a carriage ride, no amount of sorrow keeps him from filling one sheet of paper after another with letters, numbers, and drawings. Call it serene graphomania. Leibniz seems to wake up writing in the morning and go to bed writing at night. Along the way, one tower of paper rises up alongside the next, massive edifices built of pages of every size imaginable, down to the tiniest slips. Leibniz left behind about 100,000 pages, including drafts, notes, and letters. The majority of it is today housed at the state library of Lower Saxony, which bears Leibniz's name: the Gottfried Wilhelm Leibniz Bibliothek—Niedersächsische Landesbibliothek. Leibniz corresponded with about thirteen hundred people. Since 2007, the letters written by and to Leibniz have been included in UNESCO's list of documentary heritage, and it will likely take decades more for all the material to be published as part of the historical-critical edition of Leibniz's work being put out by two separate academies. He was no flyweight, then, among the great minds of world history.

Whether one encounters Leibniz while he is traveling, sitting at his desk amid his manuscripts, exchanging letters with scholars from all over Europe, or conversing with princesses and princes, it becomes immediately apparent that his fundamental attitude is consistently optimistic. But this optimism is of a very particular mold that distinguishes it from the dreamy progress-euphoria of the late eighteenth century. Occasionally, around the edges of his optimism, a certain melancholy comes into view. He has often, Leibniz writes on one occasion, thought with sorrow of all the evils we humans are subjected to, including the brevity of life, vanity, disease, and finally death, which threatens to annihilate our achievements and efforts: "These meditations put me in a melancholy mood."[5] Thus the hope that the world might be getting better seems, to some extent, to spring from the worry that things might not turn out well, that our condition might not improve after all.

In any case, it won't improve without our own efforts. Leibniz's world is the best possible only because he conceives of it as containing the possibility that humankind may strive toward the best. To continue realizing the best world anew becomes a perennial task— the daily business of a world in a permanent stage of awakening. Every day, like a small god, every human individual faces the challenge of choosing, from among many possibilities, that which is to become real. All possible worlds, according to Leibniz, strive to become real, but only one can actually exist, while all the others must linger in the realm of the merely possible. For Leibniz, who makes his ideas reality primarily by means of quill and ink, this means deciding day after day what notions he will set down on the page and what notions he will continue to develop in the process of writing. During the moment before his quill touches the white emptiness of the page, all possibilities are hanging in the air, all of them equally full of potential—but at the moment of contact, one of them enters the world.

THE

BEST

OF

ALL

POSSIBLE

WORLDS

CHAPTER 1

PARIS,
OCTOBER 29, 1675

PROGRESS, OPTIMISM, AND
RESTLESS JOURNEYING

It is not good for one to spend the whole day brooding.
Thinking is bad for one's health, Metaphysicus.
Why don't you look there at the muddy ground instead,
see what quivers and gurgles.

DURS GRÜNBEIN, *VOM SCHNEE, ODER DESCARTES*
IN DEUTSCHLAND (2003)

THE FLY

It swoops down from the ceiling and buzzes around the room, darting this way and that, fitfully changing the direction of its flight. The housefly's movements aren't so quick anymore; the days have gotten shorter, and it sinks sluggishly onto the windowsill. In the chill it grows inert, but as the stove heats up, it begins once more to zip through the small room, lit by the glow of torches and candles. By the window, there is another source of warmth. On the table next to the window are the remnants of a meal and a cup of sugar-sweetened coffee. The sweet nourishment lures the fly; it flies to the table. There the dark source of warmth begins to move. A hand raised to strike casts a shadow; quick as the wind, the insect dodges out of the way. While the hand's owner has eyes that can process only about twenty frames per second, the fly perceives nearly two hundred frames in the same amount of time. To it, the hand moves incredibly slowly, as if in slow motion; by the time it reaches the fly's position, the insect has already turned and flown off toward the stove. From there, the tiny creature observes the large warm shadow that perches in its chair for what seems like an eternity—and from time to time, it flies once more toward the tempting sugar.

COFFEE, A LITTLE WINE, AND PLENTY OF SUGAR

Bent low over the table sits a man, incessantly reading and writing, occasionally raising a hand to shoo away a fly that loves sugar as much as he does. Thus can we imagine Gottfried Wilhelm Leibniz, who has likely been in this spot working without pause since late morning. It wouldn't be hard for Leibniz to imagine what is happening from the vantage point of a fly—after all, in his conception of it, the world in its totality consists of a multitude of actors, all with different perspectives. Late summer in Paris—an especially dry one this year—is long gone; the weather is slowly getting cooler.[1] Without a heated stove, one could hardly bear to spend most of the day sitting. This, however, is the preferred way of life for the scholar from Saxony. Leibniz is twenty-nine years old, of medium height, and a bit skinny; his hair is brown. He describes himself as fairly well balanced, tending neither to impulsiveness nor to melancholy, a man with a quick wit and a lively sensibility. More than anything, he fears that prolonged sedentary study and too little movement could one day cause him an early death.[2] During the day he drinks coffee with lots of sugar, and at night just a little wine, which to the delight of the flies he also likes to sweeten a little. In France, where the Gregorian calendar is in effect, today is Tuesday, October 29, 1675. Leibniz is living on Rue Garancière—a street just 220 meters long and straight as an arrow—in the Faubourg Saint-Germain, which at this time still lies outside the city center.

On this gloomy Tuesday in late October, Leibniz for the first time writes down on paper a symbol that will fundamentally change mathematics. What flows from his quill on this Tuesday, and what is today part of the standard curriculum in advanced math classes, is a simple symbol that gathers together the knowledge of the best mathematicians of the seventeenth century in the most splendid fashion and that its inventor will further develop into the key symbol of a new mathematical method. This symbol is the $\int$, an oversize s, today known as the integral symbol; with its help, both the length of a curve and the area underneath it can be elegantly calculated and succinctly notated using one and the same calculation method. October 29 marks a high point in Leibniz's years of work on infinitesimal mathematics. In the $\int$ he creates a symbol that—as the next weeks and months will bear out—will make a crucial contribution to the development of a method of calculation that deals with infinitely small values within the framework of an easily manageable formulaic apparatus. But let's not get ahead of ourselves.

The night before this Tuesday, Leibniz probably went to bed late, as he so often does. "Stays up late and gets up later," he says of himself.[3] He is used to working tirelessly, late into the night, while others are already asleep. For about a year now, he has been living on this narrow, dark street not far from the Jardin du Luxembourg. But what has brought Leibniz to Paris? The path that led him here from distant Saxony could not have been more unusual. Born in 1646 in Leipzig in an academic household—his mother was the daughter of a renowned lawyer, his father a notary and university professor—he grew up in a time of political and religious upheaval. Broad swaths of central Europe lay in rubble and ashes. The devastating Thirty Years' War was over, but it left in its wake a fractured continent and a deeply divided Christianity split into scarcely reconcilable confessions. At the same time, the unity of faith and reason threatened to break apart as the new disciplines of rationalism and empiricism started to assert themselves. At eight years old, the bright young mind taught himself—his father had died two years earlier—Greek, Latin, and Hebrew with the help of his home library. He devoured one book after another; some of them he even learned by heart.

At first, the highly gifted boy seemed to be following in his father's footsteps. He studied philosophy and law in his hometown and later in Altdorf bei Nürnberg. But he refused the professorship offered to him after completing his doctorate and postdoctorate studies, and instead—incessantly seeking and absorbing new knowledge—he went traveling. He didn't exactly have a plan, but armed with an interest in everything and everyone, he headed toward Holland, only to end up first in Frankfurt and then in Mainz in 1668. There he managed to enter into the service of Elector and Archbishop Johann Philipp von Schönborn and took part in a major project of judicial reform. Even in these early days in Mainz, he displayed what would be his typical working method, jumping between the most disparate political, religious, and scientific issues. As a child of the Thirty Years' War and its aftermath, he sought ways to reconcile a riven Europe, to reunite the continent's Christians, and to promote wide-reaching progress in all areas of human society and civilization. Spreading a Christian faith founded on reason throughout the globe, supporting the general welfare, improving life through science and technology—throughout his life, Leibniz will feel himself duty-bound to these idealistic goals.

But how to achieve all this? Leibniz needs a universal scientific concept and many allies—like-minded scholars and, above all, powerful patrons. And ideally he will find them in one of the metropolises of Europe. Thus Leibniz doesn't hesitate when, in the spring of 1673, an opportunity presents itself for him to travel to Paris as part of a diplomatic mission. He is supposed to play a role in the attempt to keep the French government from going to war against Holland and Germany. He develops a plan for Louis XIV to lead a military operation to occupy Egypt, not only to keep the power-hungry king from launching an offensive to his immediate east, but also to place France in a strategic position that would enable her to press on and gain access to the riches of India and Southeast Asia. To this end, Leibniz even proposes the construction of a canal between the Mediterranean and the Red Sea. But the "Egyptian Plan," which anticipates the idea of the Suez Canal but never reaches the ears of the French monarch, becomes moot when, soon after its drafting, France attacks the Netherlands.[4]

Oh well, Leibniz thinks to himself, the main thing is to be in Paris, which, alongside London, is a center of science and culture in Europe. Nearly half a million people live in France's capital; Leibniz finds himself in a modern city. Countless carriages and coaches fill the streets and *allées*. Even after dark the traffic doesn't let up. Oil lamps light up the streets and are relit every evening, the world's first comprehensive street lighting in a large city. Leibniz is electrified. He dives excitedly into the whirlpool of urban modernity. Filled with drive and ambition, he seeks to enter the orbit of the social and scientific elite, above all the Académie des Sciences, where the leading scientific minds conduct research. He makes contact with several members of the academy, meeting with, among others, Pierre de Carcavy, the royal librarian, and Giovanni Domenico Cassini, who works at the Paris Observatory. He is also invited to discussions that take place at the home of the renowned theologian and philosopher Antoine Arnauld. Here he even manages to be introduced to the powerful minister of finance and economics Jean-Baptiste Colbert.

At the time, everyone in Paris is talking about the philosophy of René Descartes. Descartes's followers divide the world into a strict dualism of mind and matter; they conceive of animals as artificial automata, claiming that, despite its complexity, a living animal is not fundamentally different from a clock or a water pump. All of cultural and scientific Paris is wild about automata and machines. Their presence in daily life is increasing: there are toy automata in the playrooms of aristocratic children, siege engines in the army, musical clocks with moving figures in the courts of kings and princes, and hydraulic organs in churches. Engineers tinker with walking robots. On stage, plays are performed featuring movable machines; on the same street as Leibniz lives Alexandre de Rieux, Marquis de Sourdéac, an influential socialite and enthusiastic devotee of the mechanical theater.

Leibniz, who while in Paris will investigate Descartes's legacy, is also fascinated by this world of automata, but unlike the Cartesians, he does not conceive of nature and life as a complex mechanism. By no means does he consider consciousness and the mind to be traits exclusive to man; they are present everywhere in nature, albeit in varying degrees. This is what Leibniz means when he says the world is animated through and through. Even such tiny creatures as fleas and flies are no hollow automata but possess faculties of perception and consciousness, however simple or rudimentary. Modern neurobiologists have proved Leibniz right. Experiments with implanted electrodes demonstrate that houseflies have elementary cognitive capabilities and primitive awareness. Flies are bothersome creatures for Leibniz, too, but unlike Cartesian philosophers, he does not regard the insects that are competing for his beloved sugar as mere flying machines.

BIG CITY DREAMS

"His sleep at night is uninterrupted," Leibniz writes of himself.[5] That sleeping habit is rather atypical for the seventeenth century, when many people tend to build into a night one or several periods of waking.[6] Most Europeans don't sleep eight hours in one stretch but rather in stages. After the sleeper gets an initial three to four hours, a two-to-three-hour break usually follows, after which the sleeper turns in again and sleeps until morning. The waking periods are used for praying, for reading, for amorous diversion, or for chatting; some even leave their bedrooms to visit neighbors. Leibniz, however, sleeps through such activities, preferring to enjoy his rest and give himself over to dreams. What exactly he dreamed the night before October 29, 1675, we don't know. But dreams play an important role in his worldview. To him, they serve as further proof that the Cartesians' mechanistic conception of nature doesn't add up. Consciousness is not limited to the waking individual; rather consciousness is equally, though imperceptibly, active in the sleeping and dreaming person as well. In our sleep, we take dreams to be real, Leibniz argues in a letter to his friend, the scholar Simon Foucher, who also lives in Paris, referring to an episode concerning the dream of an oriental caliph.[7] In 1675 there is no better place in Europe than the French capital for discussing such dream tales from the land of the crescent moon. In that year the orientalist François Pétis de la Croix brings the collection of folk tales *The Thousand and One Nights*, itself written as a many-layered dream story, to Europe for the first time, and it is in Paris that he begins to translate them into French.

Leibniz's Paris years number among the most fruitful in his life: writing letters and calling on the help of his existing contacts, he systematically builds up a network of scholars that extends all over Europe. The city is an inspiration to him; he is positively bursting with ideas, working simultaneously on all manner of projects and at times reaching a veritable frenzy of inquiry. The scope of his scientific interests is practically unlimited—it is, indeed, universal. The reach of his activities encompasses—among many other things—the pondering of theological problems, like the conflict between the Catholic, Protestant, and Reformed churches over the doctrine of the Eucharist, and of philosophical questions concerning freedom, morality, and the structure of the world. Added to this is a breathtakingly wide panorama of subjects within the natural sciences and medicine: among them, experiments in physics dealing with gravity, elasticity, and statics; inquiries into anatomy and the art of healing; experiments with water clocks and the use of chemical substances or filter apparatuses to desalinate ocean water. Even curiosities like a writing sample produced by a woman with no arms, writing with her feet, stimulate his inquisitive mind.[8]

Leibniz daydreams of gaining an appointment in Paris—ideally, at the Académie des Sciences. Despite his assiduous efforts, however, which include learning to speak fluent French in just a short amount of time, as of October 29, 1675, he still has not managed to secure a position. France's capital, with its many brilliant minds, remains a forbidding environment for an ambitious young intellectual. "Paris is a place," he writes, "in which it is difficult to distinguish oneself: one finds the most capable people of our time, in all manner of sciences, and it takes a lot of work and a bit of tenacity to make a reputation for oneself."[9] He is by no means lacking in either tenacity or willingness to work.

Alongside all his studies he still has time to take part in cultural and social life. He gets as much diversion and pleasure from the plays of Molière as he does from performances at the Théâtre du Marais. Salon culture also interests him, though it's not certain whether he ever attends a meeting of the legendary Société de Samedi hosted by Madeleine de Scudéry, with whom he will later maintain a correspondence. Leibniz is no loner, even though, by his own account, he prefers reading and solitary contemplation to socializing. He describes himself in this era as being not particularly handsome, with a face that's usually pale. His cold hands and fingers, even compared to the rest of his gaunt body, are much too long, he muses. Still, he feels comfortable in society; indeed, he believes that he knows how to "entertain [a gathering] rather pleasantly" on occasion. He finds more enjoyment in "jocular and merry conversations than in games or pastimes that involve physical activity."[10] He's up for anything, so long as he doesn't have to move around too much. The beloved court game of blindman's buff must have held as little appeal for him as Baroque society dances like the courante, the gigue, or the saraband. And he knows all too well that he's not fit to play the part of hard-drinking courtier; at longer festivities, he prefers his wine well watered down.

ON THE BANKS OF THE SEINE

Wild parties at court are pleasant enough, but what Leibniz is excited about are the advances in science and technology happening all around him here in Paris. The prior month, in September, while taking a stroll along the Seine, he witnessed a man trying to walk across the river with the assistance of a flying machine. For Leibniz, this public demonstration-cum-experiment has great symbolic value. It represents the promise of being able to accomplish the heretofore impossible with the help of science and technology (namely to cross a river, half hopping, half flying). At the same time, the moment also has a religious aspect, charged as it is with the promise of enabling people to walk across water like Jesus. Given all this, the flight demonstration can be read as emblematic of a time of progress, in which inventions, machines, and new designs open the prospect of leaving existing limitations behind and soaring off into a space of unimagined possibilities.

Leibniz is caught up in this euphoria of progress. The spectacular demonstration of flight lends wings to his imagination and inspires him to dream of an academy of public presentations, with circus acts, tournaments, fencing duels, and fireworks; with stage performances that showcase anatomical exhibitions, experiments with machines, laterna magica demonstrations, marionette acts, horse ballets, simulated naval battles, and extraordinary concerts featuring rare musical instruments. There would be lottery drawings and games of chance, ball games, galleries, cabinets of curiosities and collections of natural historical artifacts, a medicinal herb garden and an office for registering inventions. Leibniz has plenty of illustrative material to fuel his academy dreams. Near his room on Rue Garancière is the colorful Saint-Germain street fair. The idea would be to encourage wealthy men in the orbit of the French king to finance this institution for research, education, and entertainment. Why not arouse people's curiosity and passions? Instead of condemning these weaknesses as vices to be combated (as the church would have it), can they not be put in the service of progress and prosperity? Leibniz doesn't deny the bad in humanity, he just wants to give it a new, positive meaning, to have it be understood as the condition for the possibility of improving the world.[11] Leibniz's optimism, which will become characteristic of his later philosophy and of the early European Enlightenment in general, is already beginning to show through while he sits on the banks of the Seine in the late summer of 1675 and follows his thoughts wherever they take him.

The Black and White Magic of Ink and Paper

But that was a few weeks ago now. When exactly Leibniz got out of bed on this particular Tuesday in late October, we can't say for certain. It can't have been too early, especially since for the past few days, he's been bothered by a cold that keeps him from going out.[12] True, he's often awake as early as six o'clock. But getting out of bed is usually the last thing on his mind. Leibniz would rather abandon himself to his thoughts. His head is already full of ideas. They come to him like forest animals venturing out into a clearing in the gray of dawn. There's hardly time to linger with one idea, the next follows so quickly. "Sometimes in the morning," he notes to himself, "in the hour that I spend still lying in bed, so many thoughts come to me that I need the whole morning, indeed sometimes the whole day or even longer, to set them down clearly in writing."[13] Usually Leibniz writes while sitting cross-legged in bed. The ideas aren't fully thought out when he sets them down on paper; rather, he formulates them in the process of writing them down. As he does, they take shape and gain clarity, which causes further chains of thought to form. As if by black magic, iron gall ink fills sheet upon sheet, the text extending right up to the margin. The letters get tinier and tinier the closer they get to the edge. Leibniz, nearsighted since childhood, keeps on writing, his face hovering close to the page.

On this late October day, it is mathematical investigations that are the main focus, inspired in part by discussions with his country-man Ehrenfried Walther von Tschirnhaus, who has been in Paris since August. Leibniz sometimes finds himself in discussion with Tschirnhaus from morning to night. While they're work-ing together, it's as if they're in a frenzy, drawing up calculations and geometric figures on a single sheet of paper. One day they met at nine in the morning. Hours later Leibniz thought a storm had come on, only to realize, when he looked at the time, that it was already night.[14] Today Leibniz is probably alone, but still his quill hardly rests. Incessantly it flies across the page. Hastily he writes down everything that occurs to him, draws and works out equations almost without pause. If he thinks he's made a mistake or miscalcu-lated something, he crosses out what he's written; again and again he starts over or adds commentary in the margins—often running perpendicular to the line of text or taking the form of insertions that sometimes resemble speech bubbles. French and Latin flow from his quill at breakneck speed, calculations are enriched with side cal-culations, curves and bounded areas complemented with tangents, lines, and points of intersection marked with numerals. Without interruption, the quill forms his thoughts. Leibniz uses every mil-limeter of the paper, which is expensive and hard to come by. Let-ters he's received from friends and fellow scholars are repurposed as scratch paper. The writing material is too precious to throw away, and today's Leibniz experts like to claim, somewhat in jest, that the ever-writing, paper-hungry scholar never owned a wastebasket. Leibniz doesn't trust his memory. He writes everything down so that he won't forget it, and he holds on to everything he writes. He would rather—as he says on more than one occasion—invent the same thing twice than nothing once.

Notes of the most varied content and subject matter are often found on one and the same sheet of paper, thoughts sitting alongside or on top of one another on the page. Next to a mathematical problem is a sketch for a physics experiment; below are excerpts from a medical treatise and a few lines that make up the first draft of a planned letter; and perhaps on the back of the page are philosophical reflections on the problem of freedom. Sometimes what leads Leibniz from one subject area to another are associations between the content of one note and that of the next; sometimes there seems to be no connection at all. Often Leibniz is so deeply immersed in the writing process that he tunes out everything around him—but the fundamental necessities of life do occasionally impinge. Once, while working on a treatise on the mechanical forces of thrust and impact, he jots down a grocery list: "Two small sausages, two chickens, four loaves of bread, and three bottles of wine."[15] A man who thinks and writes without pause mustn't forget to eat and drink.

Leibniz doesn't just develop his thoughts by writing them down, he also organizes them by cutting them up. Once he has filled up a piece of paper with enough writing, he often goes at it with a pair of scissors. Dashes of ink on the page mark the line to cut along. Whole sheets of paper get turned into strips, some of them no bigger than narrow ribbons. In this way, Leibniz attempts to separate the various thoughts by topic, eventually organizing the strips and scraps of paper systematically according to subject area. For the most part, though, he never gets around to it. New thoughts and new notes come along too quickly. Scrap after scrap, snippet after snippet, they pile up and form an ever-growing mountain of paper. His time in Paris is just the beginning. Later this flood of ink-filled pages will swell to such a deluge that Leibniz is at times at risk of drowning in it. He will spread himself too thin, be pulled in too many directions at once, and even he will describe his mess of scraps candidly as "one big chaos."[16] By the time of his death, his writing, cutting, and snippet technique will have produced thousands upon thousands of papers and bits of paper, one of the largest literary legacies of any scholar in world history. What he has left to an overwhelmed posterity is, as it was once fittingly described, "a haystack

*Snippets and strips from the collection of Leibniz's
papers at the Gottfried Wilhelm Leibniz
Bibliothek—Niedersächsiche Landesbibliothek.*

full of annals, reports, aide-mémoire, catalogs, miscellanies; a tangle of abstracts and abstracts of abstracts and abstracts of abstracts of abstracts."[17]

LEIBNIZ WAS A ROLLING STONE

This might well have been another loud morning on narrow Rue Garancière, as most of them are. At the end of the short street, work is ongoing on the Church of Saint-Sulpice, a monumental basilica. Ground was first broken almost three decades earlier, but a lack of funds has delayed construction again and again, and it's still not complete. The noise on its own might have been bearable, but combined with Leibniz's distinctly ravenous hunger,[18] it's enough to finally drive him out of bed. His late breakfast is probably a meager one, since he has little money and has to ration his food. Meanwhile, about twenty-five kilometers away at the palace in Saint-Germain-en-Laye, it's quite a different story: there Louis XIV, a notorious early riser, usually partakes of an opulent breakfast around eight a.m., with fresh fruit, pastries, and choice black tea. By comparison, Leibniz can only afford modest lodgings. There is no surviving record of where exactly on Rue Garancière he lives. Presumably it's a fairly middling, maybe even somewhat rundown rooming house. He might have a room in an establishment that, up until the previous year, was a ladies' boardinghouse, whose owner, a certain Madame Saujon, now runs it as a private pension, as she will continue to do until 1696.[19]

Leibniz has a hard time keeping his head above water. For a while he was charged with the education of a young nobleman from Mainz, Philip Wilhelm von Boineburg. By this point, however, he has all but cut ties with Mainz and is no longer beholden to his former employers from the Rhenish archbishopric. Despite the difficult living situation, he is still firmly resolved to remain on the Seine. In the spring of 1673, he turned down a lucrative offer to serve as secretary to the leading minister at the Danish court. To his friend Christian Habbaeus von Lichtenstern, who had been trying to secure the position for him, he wrote that he was not accustomed to subjecting himself to "the political caprices of this or that great lord." He preferred to keep his distance from the affairs of the courts of princes.[20] Leibniz acts self-confident and bold and considers himself a free spirit. He doesn't want to enter into the service of some prince or other and sink to the level of subaltern creature of the court, merely for the sake of a secure position. His goal isn't to serve a single nation; it's to serve humanity as a whole. He is out to promote the general welfare, the *bonum commune* of all peoples. Better to do a great deal of good among the Russians than do little among the Germans or any other European nation, he will say later in life, for he considers "heaven to be the fatherland and all well-intentioned people its citizens."[21] Leibniz resolutely insists on his independence as a scholar, bound in the end to serve only science and philosophy. And yet he seeks out those who rule. This is because he needs the support of deep-pocketed potentates to finance his scholarly work and to realize his lofty ideas and plans for human progress. The ideal would be a ruler who left him alone to study and at the same time had power, money, and the will to sponsor his ambitious projects.

Leibniz seems to have found just such a promising patron on this Tuesday in late October. John Frederick, Duke of Brunswick-Lüneburg, wants to land his services as librarian and Hofrat for a yearly salary of 400 taler, no measly sum. Leibniz has accepted, but he continues to put off actually traveling to Hanover to take up the position. Up until this past summer, he was able to scrape along by producing legal reports, even though the piecemeal payments for such unsteady work are barely enough to cover rent and food. But now he's in danger of running out of money entirely. Just a few days ago he wrote home to Saxony and asked his half-brother Johann Friedrich for financial support. To that same end, he also wrote to Christian Freiesleben, who manages the family's legal affairs.[22] While he's still waiting for an answer, suddenly a prospect opens up, and it seems possible that he might find a position in Paris after all.

Two days earlier the mathematician Gilles Personne de Roberval died. Joachim d'Alencé, secretary and adviser to Louis XIV, can imagine Leibniz taking the spot that has just opened up at the Académie des Sciences. Leibniz has gained some renown in Paris, above all for his model of a machine that would carry out arithmetical operations automatically. But he isn't satisfied with the result and continues to work feverishly on perfecting his *machina arithmetica*. And this very Tuesday, October 29, around noon, d'Alencé pens a short note urging Leibniz to hurry and finish up his work on the arithmetical machine so he can use it to apply for membership in the academy.[23]

The initiative, alas, will soon come to naught. For Leibniz, though, this isn't the end of the world—after all, he's got the position with the Hanoverian duke to fall back on. What he's been dreaming of these past few weeks and months is to be able to shuttle between

France and the Holy Roman Empire as an independent scholar, now serving one ruler, now another—or better yet, being in the employ of several princes at once. To Habbaeus von Lichtenstern, he writes that he would like more than anything to be a sort of "amphibian," living at times in Germany, at times in France.[24] What he has in mind is an autonomous life with good connections to those who wield power. A life lived perpetually on the go in the service of science, endowed with the authority and commendations of a special envoy of the highest rank. Half terrestrial, half aquatic: Leibniz's amphibian metaphor shows that for such a life, he is prepared to pay the price of not always clearly belonging. And, in fact, for his choice to live in Paris he is criticized again and again by his half-brother, who tells him he lacks patriotism toward his fatherland and is closer to the Catholic French than his Protestant countrymen.[25]

Highly mobile, hybrid careers are a widespread phenomenon in the early modern period: unofficial diplomats, spies, roving merchants, traveling performers, wandering artisans, itinerant scholars, religious refugees, unemployed mercenaries, day laborers—it's not uncommon for several of these descriptors to overlap in one and the same person. Their home is wherever they lay their head at night. They lead unsteady lives, like Anthony Standen, a notorious English double agent of the late sixteenth century who referred to his life as being in "perpetual motion and so comfortable to our English proverbe of the Rowlinge stone."[26] Leibniz is another such "rolling stone"; he can claim membership in that group of precarious persons usually deemed suspect by others because they won't let themselves be pegged as properly one thing or another. And later he too will be accused, on several occasions, of being a spy or a double agent.

Leibniz's professional and financial situation was never—neither before nor afterward—as bad as it was in October 1675. And yet as unlikely as it is, this is the period in which he makes his important breakthroughs in mathematics. To realize such an achievement at the low point of one's professional career—is it a coincidence? Not for Leibniz. To the contrary, it fits perfectly his notion that it is scarcity, not abundance, that gives rise to the good in the world. "For this I see and experience, that they who have so much often consume as much as they who have little earn. One who has less must work more. And the more one works, the more skillful one becomes. And for one who is skillful, it is not difficult to earn something."[27] Thus writes Leibniz, earning little and working hard, on October 21— eight days before he invents a new mathematical symbol, thereby more than proving his skill in the realm of advanced science. Once again an apparent negative has value: this time, instead of human vices and weaknesses inducing a positive effect, it's want and sacrifice leading to inventiveness. An idea that will become the hallmark of Leibnizian philosophy and that, in the coming years and decades, he will hone and express in ever stronger theoretical terms—namely, the idea that the bad in the world is necessary for the realization of the good—here has a concrete connection to his life: in the midst of a career setback, Leibniz lifts himself to the highest reaches of abstract mathematics and achieves the extraordinary.

THE WORLD AS FORMULA

Leibniz is hungry, hungry for knowledge and truth—but neither sates his appetite. On this October 29, however, he has enough intellectual nourishment to tune out, at least for a moment, his worries about the future and the course of his career. The day marches on, and he keeps right on reading, writing, and working out calculations. Time flies by; the hours speed past without his noticing. Dividing up the day precisely, hour by hour, isn't his thing today, though in other contexts he does value a meticulously planned schedule. When he still had the young Boineburg as his pupil in Paris, he imposed on him a strict daily regimen: wake up at 5:30 a.m, dress and pray; from 6 a.m. to 7 a.m., recap language exercises from the day before; from 7 a.m. to 8 a.m., lessons with the language teacher (with translations from Latin into French and vice versa); from 8 a.m. to 9 a.m., math lessons; from 9 a.m. to 10 a.m., mass and a sermon; from 10 a.m. to 12 p.m., lessons with the dance and fencing instructor; at noon, lunch; from 1 p.m. to 2 p.m., afternoon nap; from 2 p.m. to 4 p.m., history and geography; from 4 p.m. to 5 p.m., language lesson; from 5 p.m. to 6 p.m., guitar lesson; from 6 p.m. to 7 p.m., either reading a useful book or going to see a comedy; from 7 p.m. to 8 p.m., dinner; from 8 p.m. to 10 p.m., reviewing what was learned that day and reading a useful book.[28]

It's a good thing one doesn't have to keep to the schedules that one has stuck others with. Better to forget one's surroundings and plunge into uninterrupted study with no timetable. Today the quill flying across the page gives shape to curves, formulas, and geometric figures. Leibniz addresses himself, for example, to the arithmetical question of whether multiplication operations involving fractions can be carried out using division—and if so, how—and comes to the conclusion that such a substitution cannot be made in all cases.[29] They say brilliant mathematicians are young. That would make Leibniz, on the cusp of his thirtieth year, a late bloomer. It's true that as a student in Jena in 1663, he attended lectures in mathematics given by Erhard Weigel, but only in Paris is he initiated into the deeper mysteries of advanced mathematics. It is Christiaan Huygens, member of the Académie des Sciences and one of the leading mathematicians of his time, known above all for his studies of pendulum clocks and telescopes, who has opened Leibniz's eyes to the discipline.

Impressed by how quick a learner he is, Huygens has taken Leibniz under his wing. At his mentor's prompting Leibniz eagerly reads the relevant works on geometry, arithmetic, and algebra. He absorbs the ideas of Descartes, Blaise Pascal, and others; often, before he has even finished reading a book, he has started building on the ideas inside. Sometimes he gets ahead of himself and stops halfway. Once, apparently, Huygens actually laughs in his face. But that doesn't bother Leibniz much. He just keeps on studying, undaunted. He works his way ever deeper into several problems. Thoughts and ideas just pour out of him. Along the way, much that is quite useful vanishes into a drawer and won't resurface until the twentieth or twenty-first century, including some approaches to solving systems of linear equations, or an ingenious system of composite signs (made up of overlapping plus and minus signs) for working with common equations involving conic sections.[30] Leibniz has already come to suspect that the oft-discussed geometrical problem of squaring the circle cannot be solved with a compass and a ruler alone.[31] In the fall of 1674, he goes to Huygens with a formula for determining the value of the mysterious number pi (π). To construct the formula, he conceives of an endless series in which ever smaller fractions of 1 are subtracted and added, with subtraction and addition alternating in strict sequence: $\pi/4 = 1/1 - 1/3 + 1/5 - 1/7 + 1/9 - 1/11 \ldots$ and so on to infinity. Leibniz is neither the first nor the only one to come up with this series, but again and again in his letters and writings, he will proclaim it to be his own discovery—and with success: it will go down in the history of mathematics as the "Leibniz series."

Huygens and Leibniz alike praise the series for being particularly beautiful. Aesthetics plays a big role in mathematics. But what's beautiful isn't necessarily true. Leibniz, like others, accepts that in the search for results that are both simple and elegant, there are bound to be a few missteps along the way. Nor is elegance any guarantee of practical applicability. Isaac Newton, for example, considers Leibniz's series to be of little use in practice because it takes much too long to arrive at a value that is accurate to several decimal places.[32] In any case, with his efforts to come up with such a formula, Leibniz has stepped into the daunting arena of operations involving infinitely large or small values. What does it mean that pi has an infinite number of digits after the decimal point, and that the sequence of digits doesn't show any regularity? As Leibniz understands them, infinite numbers are not real values; rather, they are well-founded fictions and ideal elements. While it can certainly be said that something is however large or small, there is no such thing as an infinitely large or infinitely small number, because you can

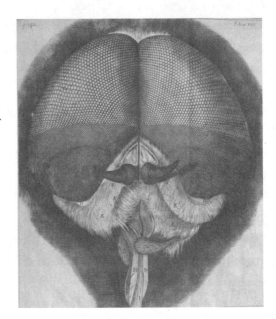

Enlarged fly eye.
Robert Hooke,
Micrographia: Or
Some Physiological
Descriptions of
Minute Bodies Made
by Magnifying
Glasses . . . *(London,*
1667), pl. 24.

always add another number to it, or subtract another number from it. For Leibniz, infinitesimal values are mere expedients that make it possible to imagine, for example, dividing up a straight line or a curve into an unlimited number of smaller sections.

Some theologians and pious individuals are leery of the new approaches in infinitesimal mathematics; to them, all this talk of infinity sure seems to contradict the conception of God's creation as finite and ordered. But Leibniz doesn't see a problem. In his view—and in line with the Bible (Book of Wisdom 11:20)—God created the world by measure, number, and weight. He characterizes the Creator as the first and supreme mathematician. In his debate with Baruch de Spinoza, whose philosophy he first encounters around this time, Leibniz will distinguish between degrees of the infinite. Divine infinity, he argues, is the highest form, comprehending all in one, in a spatial as well as a temporal respect—and as a real entity, not merely an imaginary value.[33] How, though, were the strict necessity and eternality of numbers able to fit into the inconstant and chance-ridden earthly lives of humanity? "As God calculates and puts his thoughts into action, the world comes into being."[34]

Leibniz conceives of the universe as a kind of formula. It is the task of mathematicians to set down the divine laws of creation and express them in the language of mathematics (symbols, equations, operations). As chaotic as the things of creation might seem, if we look at them more closely, we find the uniformity of mathematical structure and order. The view through the newly invented microscope seems to confirm this. In early 1673, Leibniz was able to look into Robert Hooke's device with his own eyes and marvel at the sight of the tiniest objects enlarged to many times their size. Thus he was able to see how, for example, the compound eye of a fly is made up of countless tiny shapes arranged in a regular structure that resembles a geometric pattern—exactly as Hooke had described and visually depicted it in his *Micrographia*.

SIGNATURE OF THE INFINITE

How does one calculate the gradient of a stretch of road that a carriage has to climb in order to make it from, say, northern Paris to the village of Montmartre? And how large is the total area encompassed by the Jardin du Luxembourg, whose shape is not rectangular but irregular? These questions, as different as they may seem, are related and can be answered mathematically with one and the same method of calculation. To have proved this counts among Leibniz's great accomplishments of his time in Paris. The gradient of a mountain road can be reformulated mathematically into the slope of a curve. It's long been known how to calculate the slope of a line using the formula of rise over run. But how can one use this formula to describe the slope of a curve? Quite simply, by dividing the curve into an infinite number of small lines. Suppose one stands on the beach and looks out at the ocean. The horizon may look like one long straight line, but if one reminds oneself that one is seeing only a small segment of the curved surface of the Earth, and that the entire globe is composed of innumerable tiny horizon segments, then it becomes possible for one to imagine that a curve can also be broken up into an infinite number of small straight lines.

The slope of a curve is different at every point along the curve. In order to find out what the value of the slope is at a particular point, one must find the line that touches the curve at that point. In order to find this line, called a tangent, Leibniz connects two points along the curve that are close to one another, to form another line known as a secant. Then he moves the points ever closer to one another, until the secant becomes a tangent, which, properly defined, is a line that connects two points along the curve with an infinitely small distance between them. At the same time, Leibniz develops an algebraic framework that makes it possible to use formulas to describe these geometric operations and transform them into functional equations. The whole thing can then also be reversed.

The area inscribed by a curve can be determined by dividing it into an infinite number of infinitely small rectangles, the sum of whose areas approaches the area value that one is seeking. In this operation, the tangent is likewise used as the limit of two points moving infinitely close to one another along a curve. The slope of a curve (the differential) can therefore be obtained by simply reversing the operation used to calculate the area under this curve (the integral)—and vice versa. Because in both cases the method makes use of infinitely small values (infinitesimals), Leibniz's infinitesimal calculus encompasses both differential calculus and integral calculus; the method used to calculate the slope and that used to calculate the area are inverse operations. With this inverse tangent method, a curve can thus be described as a polygonal chain, that is, as a chain made up of the sides of a polygon whose sides are infinite in number and of infinitely small length.[35]

On October 29, 1675, Leibniz is yet again focused on inverse tangent operations and continues working on his *analysis tetragonistica* (quadratic—today called integrating—analysis), which he'd begun four days earlier. Again, it's all about dividing the area under

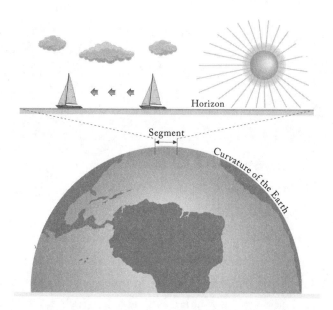

View of the horizon from the shore: segment of
the Earth's curved surface.

a curve into an infinite number of rectangles. The smaller the rectangles are, the closer the sum of their areas gets to the area one is trying to determine (what will later be called the definite integral), which forms the limit. Up to now, Leibniz has followed Bonaventura Francesco Cavalieri in referring to this sum as *omnes*. Suddenly, however, in midcalculation, he stops. He realizes that the notation he has been using so far is too convoluted and that the calculations would be easier if the notation were replaced with a simple symbol.

He hesitates for a moment, then writes: "It will be useful to write ∫ instead of *omn.*"[36] In place of *omnes*, the ∫ now steps forward—at the time, the common long form of the letter *s* used at the beginning or in the middle of a word—as a symbol for sum (*summa*), namely the sum of the areas of an infinite number of small rectangles. The width of one of these rectangles can be given as x. This x can be made to get smaller and smaller, which Leibniz expresses using the fraction x/d, where d stands for difference (*differentia*). After a few days, however, he abandons this notation and on November

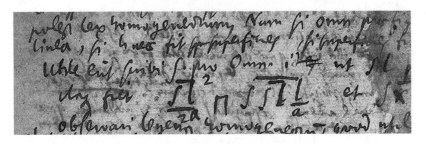

Utile erit scribi ∫ pro omn. *On October 29, 1675, for the
first time, Leibniz writes the* ∫, *the long form of* s, *later to
become known as the integral sign.*

11 replaces the fraction *x/d* with the symbol *dx*.[37] By introducing
the symbols ∫ and *dx*, he manages within the span of mere days to
lay the foundation for a formulaic language in which differentiation
and integration are treated analogously—and with that he also lays
the foundation for the modern mathematical notation used in infinitesimal calculus. The genius of this notational technique is that
it makes it possible to add together the sums of infinitesimal values using a set of simple symbols, thus enabling us to operate with
infinite values as if they were finite and in this way usefully abbrevi-
ate the long route to the calculation of the mathematically infinite.[38]

Voilà—with a stroke of his quill, Leibniz creates a symbol without which it is impossible to imagine advanced mathematics today.
On a sheet of folio paper that has only partially survived and is now
stained and faded in places, he writes, "29. Octob. 1675," which,
since he seldom dates his manuscripts, shows that he considers his
notes significant.

But what, really, is so spectacular about this? By this point, Leibniz has been aware of the fundamental concepts for a while; he has
received crucial inspiration for the development of his infinitesimal
calculus from Bonaventura Cavalieri, Honoré Fabri, Blaise Pascal,
Grégoire de Saint-Vincent, and Christiaan Huygens.[39] Yet what he
has done is invent a simple symbolic language that makes it possible, from this point forward, to concisely summarize and clearly

represent the addition of infinitely small values. In this way, he can develop a general calculation method that makes even the most difficult problems of this sort effortlessly easy to solve. Still, judging from the notes from October 29 and the following days, it's clear that Leibniz doesn't immediately abandon the old notational method. For a while, he keeps using the old designations *omn.* and *sum* or *summ.* in parallel. What we're able to see here is a gradual process of thinking and writing, of experimenting by trying out the new and old notational techniques. It's as if we can watch the "thinking by means of writing" that is characteristic of Leibniz almost in real time.

And thus we can reconstruct how, on this Tuesday in autumn, a symbol stepped out into the world that captures the mathematical infinite on paper and harnesses it to the page. The symbol kicks off an algorithm that will make the seemingly incalculable calculable, in such a way that the math will seem to solve itself and can be easily learned. In the ensuing years and decades, Leibniz and the mathematicians who will follow him, chief among them Johann Bernoulli from Basel, will keep building upon and expanding the new method. Infinitesimal mathematics provides the means for completely describing the science of mechanics, in Leibniz's time the dominant paradigm guiding both the understanding of the physical world and the technological means by which it was shaped.

Today infinitesimal calculus is fundamental knowledge for all fields that rely on being able to calculate speed, periods of elapsed time, or patterns of movement. It is necessary for the construction of curved surfaces, irregular bodies, and much more. It finds its way into nearly every area of application of even the latest technology: from the building of bridges, ships, and airplanes to smartphones, tablets, and computer-controlled robot arms. Without differential and integral calculus, physicists couldn't calculate the orbits of planets or the braking distance of cars and trains, economists couldn't follow the movement of stock prices, and doctors couldn't track the incidence of disease as it develops over the course of an epidemic.

No matter what the future has in store, technology, including artificial intelligence, will still be reliant on mathematical tools developed in the late seventeenth century.

A WORLD-HISTORICAL SLIP OF PAPER

On this October day when Leibniz sets down on paper for the
first time the now so widely used ∫, Rue Garancière is not just
loud and noisy. Foul odors abound. The streets stink of waste, the
back courtyards of urine. The rooms smell of dirty linen and damp
bedclothes. Cold sulfurous fumes rise from the stoves, and forgotten chamberpots give off a biting odor. Flies dart between excre-
ment and bits of leftover food. Prior to the eighteenth century, the
hygienic standards in the cities of Europe were low, the tolerance
threshold for foul odors high. Seventeenth-century Paris, however,
was not only the stench capital of Europe; it was also the capital of
pleasant fragrances.[40] The latest varieties of perfume arrived regularly from the city of Grasse in the South of France, gracing the
metropolis on the Seine with the scents of lavender, jasmine, and
exotic citrus fruits. Leibniz is very much amenable to these new
sensual enticements. Of his own sense of smell, he writes that he is
delighted by invigorating odors, and moreover he is convinced that
they contribute much to the refreshment of the vital spirits.[41] He
is by no means—as is often claimed—a dry rationalist who values
abstract, pure achievements of the intellect and disregards sensual
impressions. Even the long *s* is, in his understanding, a painted
thought, the expression of an intellectual discovery made perceptible to the senses. The art of mental painting becomes apparent
in the fact that with the help of a symbol like the ∫, it becomes
possible to give a typographically precise representation of that
which in geometry is expressed by graphic figures (curves, tangents, etc.).[42]

For a long time, Leibniz was accused of plagiarizing Isaac

Newton's method of fluxions in coming up with his infinitesimal calculus. But the two great mathematicians of the late seventeenth and early eighteenth centuries developed their methods of calculus independently of each other—though it wasn't until the mid-twentieth century that it became possible to prove this beyond doubt. Leibniz did conceive of his method after Newton had conceived of his, but he published it before him. Both forms of calculus revolutionized mathematics; for the first time, it was possible to carry out operations with infinitely small values in the same fundamental manner as with finite values. When Leibniz conducts his virtuosic experiments with new notation and symbols in late October and early November 1675, he is unaware that he is inventing infinitesimal calculus "only" for the "second time." But what makes his method superior to Newton's is his simple symbolic language; what he sets down on paper on October 29 demonstrates this perfectly. By comparison, Newton's long series of calculations derived from fluxions are complicated and cumbersome. Some aspects of Newton's notation have become established in physics, true, but over the long term, the symbolic language of the Leibnizian method has, for the most part, won out for use in differential and integral calculus.

How to describe, in greater detail, the successful creativity that is on display this Tuesday? Leibniz proceeds in accordance with a distinct methodical pattern: a complex concept consisting of the sum of infinitesimal values (captured by the term *omnes* or *summa*) gets expressed in condensed form by means of a simple symbol ($\int$). The symbol simplifies the concept and elevates it to a higher plane, where, in abstracted form and in accordance with certain rules, one can easily input it into further calculations. It doesn't matter which particular sum the long *s* is supposed to stand in for here. The most difficult operations featuring the use of the infinite as a mathematical limit can now be solved in a trice—as if they were being solved automatically by human reason. What up to this point could only be represented by geometrical diagrams can now be conveyed in

algebraic equations and worked out with the help of a simple calculation method. Furthermore, we can see how infinitesimal calculus
will take on an important function in Leibniz's program for a universal method of knowledge and cognition.

For what is the calculus if not a prime example of this very
program—which Leibniz had sketched out in 1666 in his treatise
on the art of combinatorial analysis (*Dissertatio de arte combinatoria*)—in action?[43] The fundamental idea is to parse the totality of
human knowledge into simple terms, to render those terms into
symbols or signs ("characters"), and then to combine them in such a
way that—with the help of an art of judgment (*ars iudicandi*) and an
art of invention (*ars inveniendi*)—the as-yet-unproven can be exam-
ined, and at the same time new knowledge can be gained. Symbolic
theory (*ars characteristica*) and combinatorics or combinatorial analysis (*ars combinatoria*) would, according to Leibniz's idea, provide
an epistemological method that would find application in all the
sciences, thus making it possible to ensure scientific progress and
develop a universal science (*scientia generalis*). Leibniz is inclined
to understand the symbolic apparatus he develops in October and
November 1675 in this sense as well. He will say so expressly a few
weeks later in a December 28 letter to Henry Oldenburg, the secretary of the Royal Society in London.[44]

Combinatory analysis and symbolic theory are meant to build on
the model of algebra, arithmetic, and geometry. In Leibniz's thinking, one can resolve extra-mathematical problems and disputes just
as if they were algebraic equations. Let us not argue anymore, he
urges; let us calculate instead: *Calculemus!*[45] In law, politics, economics, morality, religion—in all areas of human thought and action,
his program of a sign-based combinatorics could resolve conflicts,
promoting consensus and harmony. Such is Leibniz's vision. Better
to rack our brains over symbols than beat each other's brains out.
This "alphabet of human ideas," the aim of Leibnizian epistemology,[46] would naturally include the long form of the *s*.

Differential and integral calculus can thus be understood as a small part of a science-based and epistemologically grounded gospel of societal progress and global conflict resolution: promoting world peace through mathematically guaranteed truth-finding. Leibniz's conviction is that no one can hide from what is proven to be true; everyone will voluntarily submit to it. His vision sounds no less utopian today than it did back then—while he's coming up with his ∫, for instance, King Philip's War is raging in North America, one of the bloodiest colonial wars of the early modern period. But for that very reason, the vision is as timeless as ever. Indeed, in the age of "alternative facts" and "fake news," it seems, if anything, to gain relevance as a corrective ideal, in opposition to the trend of chipping away at the notion of truth. Leibniz is aware, of course, that technology and methods like infinitesimal mathematics are ideologically neutral. They can be used to build bridges, sure, but also to forge weapons to destroy those same bridges. And it seems a dangerously short step from the dictates of reason to a dictatorship in the name of the putatively rational. Nevertheless, these possibilities do not shake Leibniz's faith that there is no alternative to a calculable rationality.

At the same time it becomes apparent what the nature of Leibniz's progressive optimism is. For it is not characterized by exuberance and emotionality; to the contrary, it is sober and rational. The cheerfulness, the confident and affirmational outlook toward the future evident here, are not yet as emphatic or emotionally charged as they will be amid the eager embrace of progress at the height of the Enlightenment in the late eighteenth century. In Leibniz, optimism appears not in the bright colors of burning enthusiasm but in the plain garb of analytic rigor. The integral sign, one might say, represents a muted form of faith in progress—and thus corresponds to Leibniz's own temperament, which knows neither crushing sadness nor excessive merriment.

Today, viewing what is set down on paper against this backdrop, we can begin to understand its significance for Leibniz. We can sense how the newly created symbol fits in with his theory of knowledge and human understanding. But do we know what exactly he was thinking at the moment when the ∫ first appeared on the page? How close can we actually get to Leibniz's thoughts? Close enough to touch, or so it seems, for thanks to his exuberant passion for writing, we can reach back as far as the traces of ink he left behind in the process of reasoning with his quill. But the flash of inspiration hidden behind these traces remains invisible. And invisible it must remain, for we cannot perceive the countless tiny thought processes that led to it. They lie beneath the threshold of conscious perception—or at least, that's how Leibniz would have put it. If we follow his philosophical conception of knowledge, what we might call his psychological epistemology, such a flash of inspiration would be made up of an infinite number of tiny "flickers" of inspiration. The imperceptible thoughts that produce such a brilliant idea can be extended endlessly—just like the series of fractions in Leibniz's formula for determining the value of pi.

As Leibniz conceives of them, these unconscious thoughts consist of countless small perceptions not clearly grasped by the mind; these perceptions—infinitesimal, one might call them—together constitute a flash of inspiration like the invention of the integral sign ∫. The number of confused perceptions stretches into infinity. There is no beginning and no end; thought forms a continuum. Nothing ever happens without prior basis, and there are no leaps, whether in nature or in the mind. If these diffuse mental states could be made visible, we could stare into the unfathomable depths of human understanding. But—as Leibniz will never tire of stressing later on—no matter how deeply we are able to penetrate into the human brain, we will never descry the mind there. Traces of thoughts on paper can be rediscovered, brain waves can be measured, but the mind seems to be more than the sum of these measurable and perceptible things. It makes no difference if the mind we're talking about is that of a historical person (like Leibniz), that of a person standing right in front of us, or our own. Our perceptive faculties, as Leibniz would have put it, may be a great deal superior to those of a fly, but still a person can know the intellect or mind only in an indistinct, vague fashion, just as the fly is capable of apprehending a person only as a dark source of warmth.

No matter how simple or discerning a perception might be, it still, necessarily, requires an act of abstraction; we need it to be capable of orienting ourselves in a world of unimaginable complexity. Contemplating the mathematical infinite likewise requires that we take the shortcut that simple symbols offer us, symbols that allow the ghostly and unreal to take actual shape. To ask again: what do we see when we peek over Leibniz's shoulder on this remarkable day? Traces of ink on paper, no more than that, at least not at first, but in our minds they prompt the question of what thinking actually is and whether we can really observe someone in the act of forming (brilliant) thoughts. The difficulty, maybe even the impossibility, of a reflexive observation of the innermost depths of the mind can be given a positive spin—very much in the Leibnizian mode—because it should make us curious about Leibniz's philosophical thought, his metaphysics, which is also fundamentally concerned with such questions. And this system of thought, which will come into focus over the course of the ensuing chapters, is truly breathtaking, in many respects undiminished in its relevance, and in some respects still as provocative as it was in his time. For Leibniz's thought doesn't just assume that flies, like humans, possess a kind of consciousness, albeit one of a much lower order. In fact, according to his system, human and fly alike exist in relation to one another at the level of diffuse, infinitely small states of perception, just as everything else in the cosmos is somehow interlinked.

Evening has long since set in, the streetlamps are lit, and Leibniz likely continues working deep into the night—as he so often does. The fly that has been bugging him since morning has probably gone away. It's gotten cold in Paris, now that fall has arrived, and only a few flies manage to hang on through winter, tucking themselves away in attics or between walls.

CHAPTER 2

ZELLERFELD (IN THE HARZ MOUNTAINS), FEBRUARY 11, 1686

CREATION WITH CONCESSIONS: THE WORLD AS TASK

SILVER GALLEONS FROM VERACRUZ

Early February 1686. Several Spanish galleons lie at anchor in the well-defended port of Veracruz. The ships have ridden out the winter here in the Gulf of Mexico; their cargo holds are empty. In the harbor, the walls of the fort of San Juan de Ulúa are warm. The bright rays of the Central American sun reflect off the waters of the harbor basin. During the daytime, the temperature rarely falls below 20 degrees Celsius (by present-day measure)—to the delight of the flies buzzing around everywhere, including near the white-washed warehouses filled with ginger, pepper, nutmeg, and Chinese silk. This precious Asian bounty has been brought here from Manila by way of Acapulco. It's due to be shipped to Seville in a convoy of the New Spanish fleet, along with gold and silver from the mines of Mexico and Peru. But the sailors are still waiting for the favorable winds that will allow them to take advantage of the Florida current and the Gulf Stream.

While the galleons in the Veracruz harbor await their cargo, on the other side of the Atlantic Ocean Leibniz sits in a moderately heated room and reflects—while writing, of course—on life and the world. The town in which he finds himself lies about 560 meters above sea level in the wooded northwest of the Harz Mountain region and is called Zellerfeld. Outside, the cold is biting; the nights are long, the days mostly cloudy and dark. Inside, Leibniz is working on a summary of a long text on the fundamental principles of his philosophical worldview. What does this small town blanketed in the inhospitable gray of the Harz winter have to do with the Spanish ships in the distant port of Veracruz? Leibniz knows the answer. He happens to be situated in the most important silver mining region in Central Europe. He is here to work on projects aimed at improving the mines, but he knows that the profitability of the Harz mining operation is deeply dependent on how much Spanish silver flows into Europe from America. The price of the precious metal needed for coin production has been falling for some time now. To blame is not just a poor monetary policy but also, as Leibniz aptly judges, the surplus of the metal on the European market as a result of the "Spanish silver fleets [arriving] almost yearly from America."[1] Extrapolating from the example of American silver, Leibniz will perceive that regional circumstances are becoming increasingly interlinked with global processes.

At the moment, however, Leibniz seems rather to have his head in the clouds, working on a mysterious piece of writing that deals not with something so profane as the evolution of markets but with highly theoretical problems of philosophy. He has taken advantage of a few days in Zellerfeld when he has had little to do and is writing a philosophical treatise. The treatise is almost finished, and today, on February 11, 1686, still in Zellerfeld, he prepares a brief summary of it and sends it off in the mail, intended for two recipients. Large portions of the treatise, the summary of its theses, and the accompanying letter were composed in the wintry Harz Mountains. The notions expressed in the work don't merely reflect what Leibniz has been wrestling with for many years now on the level of theory; they are also based on tangible, concrete experiences. His abstract reflections on God, being, and substance are also an expression of what he has to contend with on a daily basis in the Harz. They show that their author, despite his high-flying ideas, has both feet planted firmly on the ground of a changing world.

Leibniz's thought is composed of overlapping layers. For him, everything is connected with everything else. This is true not only for the structure and order of the world but also for the circumstances of daily life. Leibniz communicates via networks of correspondence, and he describes the interrelation of the Harz mining operation and the global trade in precious metals. When God created the world according to the principle of achieving the greatest possible diversity through simplicity of means, he was acting economically. At the same time, part of God's creation is the task assigned to people to take part in the design of the best possible world. Leibniz's optimistic view that this might be achieved is fueled by the maxim of general optimization. He means it not just in a moral sense (deriving good out of evil) but also in a physical sense (utilizing all forces in the best possible way) and an economic one (deploying all available energy in the most efficient way). One must read between the lines of the letter of February 11, 1686, and the enclosed summary of the treatise. Doing so reveals, behind the dry and unwieldy Baroque philosophy, a meditation, still thought-provoking today, on a world constantly increasing in complexity.

THE COURT TINKERER AND
PROBLEM SOLVER

Leibniz probably barely leaves the house on this frosty February day. The winter may be somewhat milder this year than the year before, but in Zellerfeld, at its high elevation, temperatures around this time are almost always below freezing.[2] The sky is usually overcast, the view out the window gloomy. Not that this bothers Leibniz much; he turns to his desk and molds his thoughts into written form. This coming summer he will celebrate his fortieth birthday. Since his time in Paris, his hair has thinned; "a bald pate" is now showing on the crown of his head.[3] He is now a court official, and in keeping with his status, he usually wears a wig—not always a black one, as will later be the case, but also, on occasion, a *Blont paruck*—a blond wig.[4] Leibniz doesn't live in Zellerfeld year round, but he does frequently spend long stretches of time there, sometimes even several weeks or months. He has probably taken a room in a house with a slate shingle roof on what today is Bornhardtstraße; most likely he is a guest in the home of the blacksmith Lieutenant Reinhart Pfeffer, just around the corner from the Zellerfeld mint.[5] From time to time, the sound of the bells of the St. Salvatoris Church drifts over, and the occasional stroke of a hammer can be heard in the mint's forge. Otherwise it's quite quiet—especially compared to the noisy streets of Paris, where he heard horses' hooves thundering on the cobblestones day and night, and the noise of the construction of the basilica at the end of Rue Garancière.[6]

But how did Leibniz get from the metropolis on the Seine to this remote town? In early October 1676, he set out from Paris to take up his post as librarian at the court of Duke John Frederick—after a delay of more than a year and a half. Rather than take the direct route, he traveled by way of London, where he read manuscripts on mathematics authored by Newton, and from there continued on to the Netherlands, where he spoke with such renowned scholars as Baruch de Spinoza, Jan Swammerdam, and Antoni van Leeuwenhoek, finally arriving in Hanover in mid-November. The duke of the House of Welf, in his residence on the Leine River, a tributary of the Aller, was relieved. He had finally reeled the restless scholar in.

With about ten thousand residents, Hanover was a small town compared to Paris. Following his arrival in the Welf-ruled provincial duchy, Leibniz began looking after the needs of the ducal library. After a year, he rose to the position of Hofrat. In his capacity as jurist, he constantly had to pore over legal documents and draft opinions; occasionally, at the request of the duke, he penned reports on politics. Thankfully, John Frederick left him enough flexibility to pursue his scientific interests and keep up his scholarly correspondence. He loathed the daily tedium of legal work; struggling with this Sisyphean labor was the last thing the rolling stone wanted to do.[7] He used every spare minute to pursue his scholarly passions. In the realm of mathematics, for example, he busied himself with a novel form of geometry that was concerned less with points, lines, and angles and more with the position and alignment of geometrical figures in space—problems that the mathematical discipline of topology will address systematically in the nineteenth century. Leibniz has an idea to represent such spatial relations with distinct symbols or characters. In this *characteristica geometrica* or *analysis situs*, he saw the promise—similar to what he saw in the notation used in infinitesimal calculus—of contributing to the formation of a universal symbolic language.[8] At the court of Duke John Frederick, and later that of his successor Ernest Augustus, Leibniz for the most part directed his efforts toward practical inventions and useful projects for the sovereign. Among the proposals he made were: wool and silk production, fire and life insurance programs, unsinkable ships, easy-to-move cannons made of cardboard, and underwater vessels.[9] From France, Leibniz learned of the wonderful virtues of a new substance called phosphor. And from England, he learned of a newly developed steam cooker with a safety valve, penning a short piece of satire in the hopes of getting the Hanoverian duke to share his excitement about it.[10] It was the age of tinkerers and experimenters, who brought their outlandish ideas to the courts of various local lords, seeking to make their fortune. Leibniz, too, as an all-around problem solver and prince whisperer, saw in the miniature sun kings he served a thirst for power and recognition that he tried to exploit in order to put into action his progressive plans for improving human society. Pragmatically, he thought within

the bounds of prevailing economic policy. He understood himself to be acting in harmony with the fundamental principles of
cameralism, the German version of mercantilism emphasizing
a centralized economy and efficient government administration,
when he advocated for increasing the finances of the court and
countryside by means of export surpluses and import controls.[11]
"*Fürstenstaat* first," so to speak.

Leibniz's plans to advance silver mining in the Harz served this
goal. Mining revenues made up about 40 percent of the income of
the Welf dukes in Wolfenbüttel and Hanover. The main challenge
for the mining operation lay in extracting the lead ore, which contained silver, from underground while at the same time keeping the
pits dry. The water was either channeled out through a tunnel or
forced out by pumps that were powered by waterwheels. At the time
Leibniz arrived in Hanover, the Harz was experiencing a major
drought; there was hardly enough water stored up in the retaining
ponds to drive the waterwheels and keep the pumps in operation.
Not long after taking up his position, he came up with the idea of
using windmills as an alternative way to power the pumps. The duke
commissioned him to develop wind power for the mining operation,
and to that end, he began traveling to the Harz regularly. Between
1680 and 1686, he made more than thirty trips to the northern
German highland region. During these years the Harz became his
second home. He usually stayed in Clausthal or Zellerfeld, where
the mining authorities were based.

TILTING AT WINDMILLS

As is again the case on February 11, 1686. Leibniz has been back
in Zellerfeld since early January. It's Monday, early in the morning;
in Clausthal, the next town over, the hard work in the ore mines
has already begun. It's fair to assume that he's not in any hurry to
get up right away. Usually a servant comes in to heat the stove first.

Leibniz has been making trips out to the Harz for more than five years now, but his projects haven't really gone anywhere yet; they keep sputtering out. The local mining officials view Leibniz, who holds his own shares in the mining business, as a competitor. Leibniz is tinkering with technologies that put their jobs at risk. He pays no heed to the toxic lead fumes that escape from the furnaces where the metal ore is smelted.[12] Often he refuses to descend into the tunnels to see with his own eyes that something doesn't work the way he imagines it does. What he wants is often unclear; the drawings of his complicated contraptions are often just rough sketches. The locals view tall and gaunt Leibniz, the "intellectual" from distant Hanover, as arrogant. He ignores most of their suggestions. Soon enough he acquires a reputation as a dangerous man deserving of ill-treatment.[13]

At the same time, Leibniz is struggling with additional difficulties. The Little Ice Age is causing particularly low temperatures. Ice, snow, and steady subzero temperatures take their toll on workers and equipment alike. Again and again he hears complaints about frozen lubricant or iron that has turned brittle in the cold. Meanwhile back in Hanover, the court's patience seems to be running out: the powers that be are threatening to stop paying for Leibniz's wind power experiments. Instead of windmills whose sails spin vertically, he is now putting his hopes in windmills with horizontal sails. The idea is for the mills to pump water to ponds and pits located at a higher elevation; from there it can flow down and move the wheels that power the pumps. But there are all sorts of problems. In December 1685, the wind-powered Archimedes screw that is supposed to draw the water up into the retaining pond springs a leak. And to make matters worse, the people he counts on to keep him informed are proving less than reliable. Jobst Dietrich Brandshagen, who supervises the efforts in the mines in Leibniz's absence, comes off at times as chaotic and disorganized. Shortly before Christmas, he misses several days and is unable to monitor the test runs in Clausthal because he doesn't have a rain jacket and coat with him.[14] Who would guess

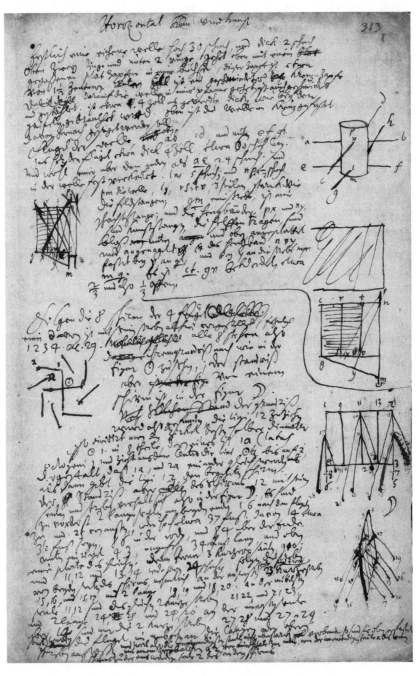

"Horizontal wind power": Notes and sketches by Leibniz
for the construction of windmills with horizontal sails
for the Harz mines, c. 1684.

that one might need weather-appropriate clothing high up in the Harz Mountains in December?

Dry summers, cold winters, constant friction with local mining officials, overwhelmed assistants, and incessant equipment damage: Leibniz's struggle with his windmills seems no less epic than Don Quixote's. *What's the use*, he might be thinking to himself this February 11—and not for the first time. The funding from Hanover has long since dried up. Leibniz still has permission to continue his mining experiments, but from now on he has to foot the full cost himself. He is experimenting with a whole host of additional technologies, among them a special iron chain (*Endloskette* or "looped chain") that makes it possible to apply counterweight while extracting ore. Master carpenter Hans Linsen is tasked with supervising the efforts at the Thurm Rosenhof mine and with paying the miners. Equipment and labor costs are devouring ungodly sums. Wet weather and sub-zero temperatures make the work more difficult; the iron chain keeps breaking or getting snagged somewhere because the shafts leading down into the mine aren't perfectly vertical. And again and again the miners complain to the mining authority that Leibniz's experiments are getting in the way of the daily work of extracting ore.[15] Coping with complaints and calamities all around, experiencing setback upon setback, feeling enormous pressure to succeed—yet Leibniz remains persistent, lets nothing unsettle him. Giving up is not an option.

At his base in Zellerfeld, Leibniz receives an update from Linsen, the master carpenter, about once a week. The short notes are brought by couriers, probably *Pochknaben*, young boys who help to break up chunks of ore into smaller pieces and often run errands for tips.[16] The latest note came two days ago, on Saturday, this time with just a curt request from Linsen asking for an additional three taler for overseeing the work in Clausthal.[17] Leibniz knows that he won't hear from Linsen again for another week or so. This gives him a few days' respite from the tough, time-consuming business of his mining ventures. He is more than grateful for these intervals—and knows how to take full advantage of them.

White Snow, Black Ink

This Monday is no exception. The pile of unfinished business is high. Leibniz has brought plenty of paper to Zellerfeld. Some of the sheets are still blank, others completely or partially covered with writing, crisscrossing the page, with lots of cross-outs, as usual. There must be dozens of loose sheets of paper in the room. Even far away from his regular desk in Hanover, he quickly fills page after page. Among all this writing are passages and excerpts copied out from historical chronicles and records. Last summer, with his projects in the Harz going worse and worse, his duke had commissioned him to write a historical work on the Welf dynasty. Along with the historical material are naturalist notes and drawings that he made on his last excursion to the Harz in the fall and that examine things like bones found in caves and fossils discovered in shale. Another sheaf of papers has almost no writing on it yet; Leibniz is planning to use it for a lengthy theological study in which he will advocate for the reunification of the churches.[18] These folio pages bear the same watermark as those of another manuscript that is covered in cramped script running to the edge of the page.[19]

That manuscript is a heavily reworked draft consisting of twelve folio pages with writing on both sides. Written in French, it bears no title and begins with the words "God is an absolutely perfect being." Immediately Leibniz starts over and writes above this line, "The idea of God," only to change it again: "The notion of God."[20] He continues writing from here, crosses a few words out, adds others, starts over yet again . . . and so on. Again and again it becomes apparent how much Leibniz is struggling to form and shape his thoughts, how much effort he puts into trying to find the right words and turns of phrase. Only in the quiet periods between his reports on the work in the mines is he able to concentrate on his writing. Now the handwritten treatise is well advanced. It holds a great deal of significance for him—after all, in this text he intends to bring together his fundamental philosophical observations and reflections of the past decades.

What exactly is in this manuscript? It deals with questions concerning the existence of God and the makeup of the world, the mystery of human freedom, the origin of evil, and salvation through Jesus Christ. It deals, in other words, with several major contentious issues of Christian philosophy in the Western tradition. Leibniz is searching for a philosophical foundation upon which the hopelessly feud-riven church can be reunited. He is convinced that the various doctrines can be brought back together on the basis of a rational philosophy. What Leibniz is attempting is to unify faith and reason. It's no small task, but as he has already shown with his search for a universal method for establishing truth, he is not daunted by the big picture; he wants to get to the bottom of everything. He is brave enough to take on such fundamental questions, but not everything is fully worked out yet; much of it is still probing and speculative. He needs to test the impact of his ideas on readers. But on whom can he impose the task of wrestling with so many new and difficult propositions? Leibniz knows that in his reflections there is much that is difficult to digest and potentially explosive.

A THREE-WAY CONVERSATION

Today, February 11, Leibniz dares to take the next step and introduce others to what he's written. The draft isn't finished yet; all Leibniz has so far is a partial copy, written in the hand of a copyist, that he is still correcting. Yet he has reached the decision to at least send a summary of the treatise by mail to two confidants. The intended recipient of the letter he plans to send today is the theologian and philosopher Antoine Arnaud, whom Leibniz got to know in person in Paris. In France, Arnaud is considered a major authority on theological and philosophical matters. He made a name for himself as an early critic of Cartesian philosophy, as a proponent of reform within the Catholic faith, as a leader of the Jansenist movement, and as a co-author of the *Port-Royal Logic*. Arnauld is not just a brilliant theologian and philosopher, he is also an expert in mathematics and logic. Just the right mix, then, to be able to appreciate Leibniz's latest effort. The problem is, Leibniz doesn't know where the seventy-four-year-old scholar is living. For many years now, Arnauld has been in hiding, moving between towns in the Spanish Netherlands (present-day Flanders). In 1679 he had to leave Paris, having taken the side of the Roman Curia in the dispute between Pope Innocent XI and Louis XIV over the matter of *droit de régale*.

One of the few people who knows where Arnauld is currently living is Landgrave Ernst von Hessen-Rheinfels. Leibniz has been corresponding with him since 1680. The two of them maintain a relationship marked by mutual confidence; both argue in their letters for a reunification of the Christian confessions. Their outspoken correspondence gives the impression that they live in an atmosphere of religious liberty and freedom of thought—but this is not the case. If any one of their letters were to fall into the wrong hands, it could cost the sender or the recipient his life. In times of rigid confessional division, religious tolerance requires substantial courage. On at least one occasion, the two men have met in secret in northern Hesse, to speak freely about a rapprochement between the hostile churches. Again and again the Hessian ruler and convert to the Catholic faith tries to get his Protestant friend Leibniz to come over to the other side. And again and again, when Leibniz refuses, he accepts it. One cannot change one's convictions at will, Leibniz says; they are founded not on volition but on a basis of reason.[21] In turn, Leibniz patiently listens when his correspondent regales him with accounts of his rakish escapades with young courtesans on his regular trips to Venice. Leibniz is rather inexperienced in such things.

BEYOND PHYSICS

So Leibniz sets about preparing the letter he's going to send. By now, the day is well advanced. On the frozen Lower Eschenbacher Pond, about a fifteen-minute walk away, children are probably skating or playing an early form of ice hockey. Leibniz has summarized the lengthy rough draft of his discourse in thirty-seven short theses. Why this condensed form? It must be entirely clear to Leibniz that much of his summary will come off as unintelligible. Maybe he intends to send a fair copy of the complete text later on and is hoping for a quick reply that he can incorporate into the manuscript before then. This would be a typical example of his dialogic approach, involving those with whom he shares his writing in the writing process as early as possible. He begins his accompanying letter to the landgrave by saying that during a few idle days, he has penned a little text.

He's downplaying things a bit. In his letters from the Harz, Leibniz likes to make a point of mentioning every now and again that he doesn't have access to his library there, thus to leave the recipient of his letter all the more impressed, for his writings even from the remote outpost of Zellerfeld are bursting with erudition. Leibniz finishes his letter to the landgrave and encloses the summary with the request that he forward both to Arnauld. Before sending it all off, he produces an abstract of the letter and enclosed text for his records—and fortunately so, since the materials he sent have not survived, while the "Extrait" has, dated "1/11 Febr. 1686," in accordance with both the Julian and Gregorian calendars.[22] For various reasons the dates Leibniz assigns to his papers are not always reliable; sometimes they're even deliberately manipulated. In this case, however, he makes sure to note the important date for his own records.

In his letter to the landgrave, Leibniz refers to his untitled work as a small discourse on metaphysics (*Un petit discours de Metaphysique*).

Later scholars on Leibniz will commonly title the treatise *Discourse on Metaphysics* and insert the theses sent to Arnauld care of Landgrave Ernst into the text as subheadings for the individual sections.[23] The word *metaphysics* here doesn't mean magic or mysticism; Leibniz rejects both, judging them to be largely attributable to superstition (though he still keeps an eye out for anything that could be put to scientific use). Originally scholars involved in organizing Aristotle's works used the term to refer to those they placed after (*meta*) his *Physics*. Later the term, first used in this bibliographic sense, evolved to take on a new meaning, as a label for the philosophical discipline that interrogates the world of phenomena, asking what is its actual basis. Metaphysics deals with questions of ultimate justification that cannot be derived from what we humans are capable of perceiving with our senses. Beginning with the writings of Immanuel Kant, if not sooner, such questions became problematic, because they were found to lead reason beyond its own limits. Ultimately they are questions that people will probably always ask as long as they are not satisfied that the world is as it appears to be. For Leibniz, however, metaphysics also means an incessant process of inquiry that refuses to be satisfied with superficial explanations and instead seeks to drill down to the root of things.

Still, today there is a widespread conception that the arduous business of the Harz mines, which dragged on torturously for many years, continually kept Leibniz from his attempt to found his own metaphysics, his true chief desire as a philosopher and universal thinker. How is one supposed to concentrate on profound meditations on the true, real world that underlies earthly phenomena when one has to contend on a near-daily basis with mundane headaches like frozen iron chains, becalmed windmills, and obstinate miners? But this impression is misleading. Things, as Leibniz understands them, are connected; everything is related to everything else. There is no divide between the everyday world we experience and the metaphysical world in the background. His technical experiments and economic initiatives are part of an all-encompassing whole.

His work on the Harz mining operation not only reflects Leibniz's understanding of how the world is structured, how it functions in its innermost core, and how it will likely continue to develop; it is also his way of participating in forming and shaping world events. Seen in this light, the Harz project, far from keeping him from further developing his metaphysical work, instead helped him to do so.

POSSIBLE ADAMS, POSSIBLE WORLDS

Winter's iron claws hold the day fast in its grip, but the room is filled with warmth. Wooden logs crackle in the stove, a sound likely complemented by the occasional patter of mice feet scurrying on the wooden floorboards. Zellerfeld, about one-third the size of Hanover, is still recovering from the devastating fire that laid waste to the town fourteen years earlier: in 1672, more than two-thirds of the town's 550-plus buildings burned down. In the past few weeks, many residents have also been complaining of coughs and runny noses. Leibniz isn't much bothered by the wet, wintry season; he rarely suffers from colds.[24] He knows that it will be a while before he receives a reply from Arnauld to the letter he's written today. When the older man receives it, he will probably think back to when he met the young Leibniz in Paris. At the time he considered the German Protestant to be a promising scholar who, alas, happened to belong to the wrong faith. Now, however, Arnauld is horrified at what has reached him from distant Zellerfeld. He is angry, and instead of responding to all the theses, as Leibniz hoped, he gets stuck on a single one, Thesis 13.

Leibniz has claimed here that every individual contains once and for all everything that will ever happen to them. At the level of term logic, as first developed by Aristotle, this means that a concept like "Adam" contains all its predicates within itself.[25] Arnauld has an inkling of where this proposition could lead. In his reply to Landgrave Ernst on March 13, he writes that he sees in Leibniz's thesis the danger of a pervasive determinism. If God, in deciding to create Adam, at the same time has decided on a particular world and all the events fated to proceed from it, then all these things must happen by necessity. If this is the case, then neither God nor the individual would have the freedom to act any differently or to influence the course of events. If Adam's fall from grace was inherent within him from the beginning, then he cannot be held responsible for it.[26]

Leibniz is horrified by Arnauld's blunt response. Still, he might have expected as much, having learned from the landgrave that Arnauld could occasionally react in a heated and quick-tempered manner.[27] Aggrieved, Leibniz replies: from among an infinite number of possible Adams (*Adams possibles*), God chose a particular one, because only this one Adam best corresponded to God's chosen order of the world. In this Leibniz sees nothing that contradicts either the freedom of God or the freedom of created minds (*esprits creés*).[28] The gauntlet has been taken up. Thus begins a passionate philosophical controversy between the two scholars that will continue until the spring of 1690. More than anything, Leibniz uses the correspondence to hone his ideas. Begun on February 11, 1686, the exchange will become a way of testing his metaphysical system of thought, which grows increasingly distinct as the debate proceeds. More than anything, what will take shape is Leibniz's supposition that God could have created any one of an infinite number of possible worlds but chose only one to actually create, namely the best one.

JUST ONE BEST WORLD

Leibniz characterizes the world as the best state or best republic. From this, the concept of the "best of all possible worlds" will later be derived, even if Leibniz probably never used this exact wording himself. The reflections connected with the idea nevertheless lead to the heart of his metaphysics.[29] God can't create what is best for every single thing; rather, he has to take care that all the parts and processes of the framework of the world are in agreement and compatible (*compossible*) with one another. To do this, God must accept compromises. Trees don't grow all the way up to heaven; all creatures are limited in their perfection by the prevailing conditions of the world around them. Everything must fit together. If God had created the world in the most perfect manner, he would merely have reproduced himself, because the highest perfection is an attribute of him alone. Leibniz's God doesn't act arbitrarily; rather, he is bound to the eternal truths of reason. Among these Leibniz counts some general principles of logic: for example, the Law of Contradiction (a statement and its opposite cannot both be true at the same time) and the Principle of Sufficient Reason (no fact is true and no statement is accurate without there being a reason why things are thus and not otherwise). Where Arnaud assumes a God who for the most part acts in secret,[30] for Leibniz divine action is transparent and comprehensible. In this respect, Leibniz consciously sets himself apart from the prevailing conceptions of God in the Christian theological tradition.

In the world as it actually exists, not everything is the way it is by necessity. Alongside the eternal truths of reason, there are also truths of fact. Leibniz refers to the latter as contingent, because their opposite is also possible. But they are distinct from the merely possible in that they actually exist as a matter of fact. By contrast, the concept of the possible also includes everything that is conceivable but doesn't necessarily have to happen. Accordingly, Leibniz attaches to the extant world a retinue of countless possible worlds that never become reality. (Parallel worlds are barred from Leibniz's metaphysics.) All possible worlds press toward existence, but only the world in which everything is optimally aligned in relation to everything else, in which everything fits together, becomes reality. This also includes evils like natural disasters and moral failings. They are necessary in the grand scheme of things in order to bring forth the good—Adam's fall and Judas's betrayal included. Without

sin, there can be no striving toward progress. Only through the failings and sufferings of human individuals does society move forward. In this way, every evil is paid back with interest, so that when
everything is added up, there is more perfection overall than there
would be if all the bad things hadn't happened.[31] Thus, in Leibniz's metaphysics, Adam's fall and its aftermath are consequently
turned into something positive. Here Leibniz takes up the concept
of the "happy sin" (*felix culpa*), which in the eighteenth century will
become central to the Enlightenment's philosophy of history.

At the same time, this world contains many possibilities. The
world is the best not simply in its presently existing state; to phrase
it this way is to truncate Leibniz's notion, as will often be done later
on—for example, in Voltaire's *Candide*. In speaking of the best of
all possible worlds, Leibniz intended not a mere description of what
we find before us; we are also to understand it in a normative sense.
What has happened can no longer be changed. But, says Leibniz,
one shouldn't await the future idly, with arms crossed. Rather, we
are called to act in accordance with the presumptive will of God,
to work deliberately toward the goal of perfecting the world—or in
concrete terms, to contribute to an increase in the general welfare.[32]
For Leibniz, the realization of the best possible world becomes a
task that challenges humankind to help shape its environment and
to work on its own (above all moral) self-optimization.

The world is always already the best because it carries within
itself the possibility of its optimization. The human individual is an
indispensable part of the best possible world's potential for realization. What is crucial is not the best per se but rather the path that
leads to it. Creation's potential to discover and to exhaust all possibilities, to strive ever anew toward the best thing imaginable, thus
becomes a spur to humanity, a purpose to human existence. And
it is precisely here that the Harz comes into play. At some point,
a pastor in the eastern Harz, the "Rev. M. Eicholz," declares in a
sermon that the use of wind power in mining is "impossible": it was
God's will to make windmills unsuitable for mining (in Leibniz's

polemical rendering of the pastor's words); otherwise God "would surely have built windmills in the Harz Himself."[33]

Leibniz sharply rejects this kind of theological critique of technology. His understanding of technology and practical science rests on the notion that man is willed by God to participate in the completion of creation by striving toward progress and improvement. The fact that something doesn't succeed right away hardly indicates that God determined its failure. For Leibniz—who despite many setbacks never despairs—such an assumption would simply be false metaphysics. For him, science and technology are by no means at odds with creation. To the contrary, they represent a chance to push the door to the possible open ever wider and to advance ever deeper into the realm of what heretofore has seemed impossible, even if it will never be the case that everything is possible. At the same time, what is inherent within the Leibnizian concept of the best of all possible worlds is the salvational promise of an ongoing labor on the possible. For Leibniz—this optimistic scientist, inventor, and project leader—nothing is impossible. He constantly sounds out and recasts the limits of the doable.

THE CONNECTEDNESS OF THINGS

The different strands of thought that Leibniz sketches out also make it clear, again, that he understands the world as a network of interconnected events. The best possible is never to be regarded in isolation; rather, it should always be seen in relation to everything else. Leibniz's metaphysics, as set down on paper in these quiet winter days in the Harz, is very much a philosophy of reticulation. The world is an interwoven phenomenon; all its parts and processes are interlinked; they mesh and complement one another in an optimal way. Arthur Schopenhauer will later make the cutting remark that ours is the worst of all possible worlds. If it were just a little bit worse, it wouldn't exist at all. But Schopenhauer's claim still rests on the same assumption, even if the outlook is reversed: the given world exists not as an assemblage of parts but rather as an interconnected whole.

For Leibniz, the world is structured according to an economic principle that is as simple as it is fundamental. The simplicity of means is balanced against the wealth of effects, as he notes in Thesis 5.[34] Herein consists divine perfection. Of all the possible worlds, the one that is realized is that in which the maxim of efficiency is best implemented, the world in which the greatest results are produced with the least effort. In the combinatorial game of infinite possibilities, the ones that are selected are those that, working in concert, produce the highest density of relations and cohesive force. The resultant world is the simplest and at the same time the richest; in it, the greatest diversity results in unity, just as a sphere concentrates the greatest possible volume in the smallest possible space. At the same time, the economy of the world is constituted in such a way that nothing in it goes to waste; every input of energy is rewarded with gain and every harm compensated with utility. Absent additional intervention from without, the world functions like a perfect machine in which all the components fit together seamlessly and any adjustment to one part has an effect on the totality of all other parts. It's no accident that in this context, Leibniz compares the creator to an expert mechanic or engineer (*habile machiniste*) whose machine achieves the greatest result in the simplest manner.[35]

Here Leibniz's metaphysical reflections begin to overlap with his ideas as a mining engineer in the Harz. His efforts in that capacity aren't directed solely toward introducing individual mechanical innovations. His intention is to convert the entire mining operation into a self-contained, technologically interconnected system that relies on the linked interplay of water power and wind power. Wind power alone, as he has come to realize, is not sufficient to power the pumps that drain the water from the mines or to power the crushing mill that breaks up the ore. Yet inconsistent wind energy can be employed more effectively if it is used to help maintain a consistent

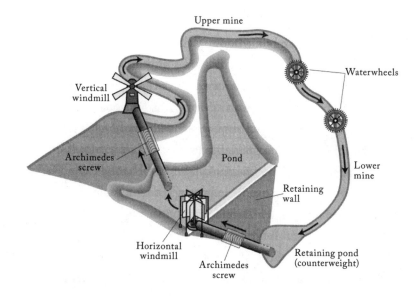

Energy system and technological array:
Planned water circuit with windmills.

flow of water. A mix of wind and water energy should spare animal and human resources and make the mining operation less dependent on particular weather conditions.

To his model of a system of water and wind power, Leibniz adds a reversible waterwheel mechanism, which should make it possible to move the ore barrels up and down the shaft, depending on the direction of the wheel's rotation.[36] And that's not all. In the very weeks in which Leibniz is composing his *Discourse on Metaphysics*, he is tinkering with a self-regulating mechanism that will cause the sails of the windmill to rotate in the direction of the prevailing wind. More experiments will follow in the years to come: for instance, with a conical cable drum and a winding drum that make it possible to achieve an almost complete counterbalance when hauling ore to the surface. Finally, Leibniz will also develop creative ideas for salvaging materials. He ponders, for example, how to recapture metal from the smoke given off by the smelter, and whether he could have gears for his calculating machine made at the Clausthal mint.[37]

What Leibniz envisions could be described as an integrated system of continually circulating energy.[38] In his mind's eye, he sees a model technical mini-world, with pumps, waterwheels, tunnels, and shafts, with windmills and Archimedes screws, with retaining ponds and streams, with cable drums and hoisting chains—a microcosm of ideal physical processes and counterbalancing forces, based on the principles of incorporating all available resources, compensating for variable conditions (like wind and rain), instituting self-regulating control mechanisms, reducing expenditure of effort, and establishing regular, stable procedures. Evident in this undertaking are all the fundamentals that hold true for the constitution of the world as a whole. Leibniz's plan, in his attempt to improve the reality of one region of the world, reflects, in his view, the reality of the globe, which is so constituted as to move toward constant optimization—and also serves as an excellent example of how the individual (here, the clever inventor and designer of machines) can play a productive role in the continual shaping of the best of all possible worlds.

EVERYTHING IN MOTION

One may see how much technology and philosophy, physics and metaphysics, are interconnected for Leibniz when one considers his observations on force and motion. Lengthy passages in his *Discourse on Metaphysics* are devoted to Cartesian mechanics. Landgrave Ernst and Arnauld are able to read only the heavily abbreviated form of these remarks in Theses 17 and 18. The two of them can only guess at what Leibniz might have meant. His position becomes clearer in an essay that he writes in Zellerfeld that appears a few weeks later in the March edition of *Acta eruditorum* under the title "Brevis demonstratio erroris memorabilis Cartesii" (Brief Demonstration of a Notable Error of Descartes).[39] The article deals with the question of how to calculate the force of a body in motion (for example, a swinging pendulum or a falling rock). In the Cartesian view, force is equal to mass times velocity (today rendered as mv). But Leibniz defines force as the product of mass times velocity squared (mv^2). Here he comes close to what will later be known as kinetic energy. In his time, however, Leibniz's immediate reward is to bring the entire Cartesian community down on his own head, which leads to a long dispute carried out in the pages of *Acta eruditorum*.

Leibniz is convinced that force in the universe is never lost but rather is perpetually transferred. His blueprint for an integrated system of wind and water power in the Harz rests on this assumption regarding the perpetual transfer of energy. For him, everything is in constant motion. He conceives of motion as not a temporary but a permanent condition of all mobile bodies. Even the state of rest is just a state of infinitely slow motion. Leibniz almost seems to grant physical bodies just as little rest as he does himself; he will later speak of mental calm as a preliminary step on the way to stupidity. It's not surprising then if, in this world of perpetual unrest, in which everything is interwoven with everything else to form a coherent whole, every little movement continues on, like an oscillation, into the most remote distance. The effect of every single action and every single event stretches into infinity, in both space and time. What is absent is just as connected to what is present as the future is to the present moment and the past. In these endless chains of causality, in this tangle of overlapping sequences of cause and effect, the slightest, unnoticed stimulus can usher in the most powerful changes.

Now, unexpectedly, a tiny old acquaintance joins our discussion, an uninvited guest whom we first met in Paris and who keeps darting every which way through the life and thought of Leibniz, always providing creative disruption: the fly. "I like to say that a fly can change the course of the whole state, if once it buzzes past the nose of a great king while he is caught up in important deliberations; for while it may be the case that he is, shall we say, weighing two things in his mind, and that there are strong reasons found on both sides, so can it occur that those suggestions carry the day which his mind dwells on longest, and this the fly can do, it can impede and disturb him when he wishes to properly regard something else, so that accordingly it never quite regains his sympathy."[40]

As suddenly as the fly appears, it vanishes once more. Tiny occurrences can tip the scales and steer the course of events in an unexpected direction. Thus in the dark weeks of the Zellerfeld winter, Leibniz develops the basic framework for his later work on dynamics. In so far as it is conceived of as an alternative to mechanistic natural philosophy, *meta*-physics in this context also means going beyond the physics of the Cartesians.

GLOBAL IMPROVEMENT AS ADMINISTRATIVE PROJECT

Leibniz's world is conceived from the top down. It is not built from the ground up, block by block, as in the biblical creation story; rather, God has recognized it in advance as the best one (including all potentialities and everything that is to come) and thus made it reality. So it is only logical that God is characterized in absolutist terminology as the supreme prince or monarch, as "the best of all possible rulers."[41] This anthropomorphic analogy is of course also directed at the princes themselves, who are meant to draw an example from God's infinite goodness and justice. Leibniz's plans for the Harz are likewise addressed first and foremost to the prince he serves. The optimization of the Harz-world is from the outset intended as a mercantilist or cameralist project. It's not just about making mechanical or technological improvements; it's also about industrial management and business administration, geared toward the prince's top-down governance.

And so Leibniz has proposed to Duke Ernest Augustus that data be compiled on the topographical features of the mining regions, as well as on the location and extent of the water sources, mines, retaining ponds, and lumber supply. In addition, maps should be made that show a "scenographic or perspectival survey of the mine," plus mine and pit tables compiled at weekly, monthly, quarterly, and yearly intervals; timesheets; reports on the crushing mill and smelting works; minutes from meetings of the mining authority; and lastly, detailed accounting statements relating to the mineworkers, foremen, and suppliers. All of it could then be recorded in the form of a bookkeeping summary on a brief "main sheet," which would give the duke and his agents a quick comprehensive overview.[42]

The weblike interconnection of real conditions, the composite system of water and wind power plus pump and mining technologies, should be enriched and enhanced by means of an organizational system of data and informational links, recorded and presented in the form of written reports, tables, graphs, diagrams, and models. Mining operations should be organized and directed by a modern administrative office.[43] Leibniz recommends himself to the Hanoverian court not only as a leading engineer but also as the duke's on-site representative, the cameralist director of a project for perfecting the local reality in order to increase state revenues and improve the welfare of all subjects. It is, in short, a project geared toward the realization of the "best of all possible Harzes."[44]

Peering out from behind the arid words of philosophical abstraction is a vibrant, vivid, variegated world, with complex processes and a comprehensive system for managing information—and with all these ideas, and only so much time, Leibniz hopelessly overextends himself. On this gray February day, while he is busy condensing his metaphysical thesis for the landgrave and Arnauld in this small mining town, back at the court in Hanover hardly anyone still believes that the Hofrat's ambitious projects for the Upper Harz silver mines can actually succeed. Up until December 1686, Leibniz will keep on coming back to the Harz for a few weeks at a time, but as he's here working on his *Discourse on Metaphysics*, he is also in the process of winding down his affairs. The trials with the looped cable have been discontinued, and the horizontal windmill and the water screw have been dismantled. He has to submit a meticulous accounting of all of it. It's a time of stock-taking, in which Leibniz sets down his thoughts on freedom, causality, and God's choice of the best. Everything is painstakingly recorded. Today at the state archive of Lower Saxony in Hanover are numerous tables and lists that have survived from this time. In them is recorded and precisely documented every individual component used for construction of the windmills, every axle-stabilizing wedge, every wooden pole, every iron ring, indeed every single nail.[45] Instead of guiding the duke, Leibniz has to answer to the court.

Between 1692 and 1696, Leibniz will again experiment with new technology in the Harz. Many of his inventions, such as the conical cable drum, will be adopted or will reappear in similar designs thought up by others independently of him; they will help make possible the array of technology in use today.[46] With his ideas on engineering, economics, and information-gathering in politics, Leibniz has run up against the limits of what is possible in the preindustrial age. Through his correspondents in England, he indirectly witnesses initial efforts to use steam engines in coal mining. But coal was not available in the Harz, so until the early nineteenth century, classic waterwheels were employed to power the pumps there. Still, failure—to go by Leibniz's own assessment of his personal contribution to the future—is not the opposite of progress. Progress requires failure; failure is a necessary component of progress. Trial and error: there is no success without mistakes. Leibniz's inventions and blueprints for projects, seen according to his worldview, are part of a merely possible world and must remain on the margins outside the real one.

The world is full of possibilities, but not everything that is possible gets to become reality. Possibilities jockey with one another in competition for the chance to exist. Perhaps this idea of competition is a contribution that Leibniz makes to the intellectual history of capitalism.[47] In any case, the idea encourages Leibniz himself, spurring him on, helping him to endure setbacks and to persist in trying again, and on the other hand, allowing him to accept with equanimity when something—be it a method of calculation, a technical invention, or a model for organizing knowledge—doesn't work out. If that happens, then it just wasn't meant to be—in this world, anyway, or at least not yet. Things must fit together—otherwise they make no sense. Leibniz is prepared to accept the verdict of incompatibility, or incompossibility. This is what distinguishes his notion of possible worlds from that of a world of unlimited possibilities. Leibniz's concept of the world does indeed value the concept of the possible, but not in the manner of a modern age that thinks in terms of endless possibilities. Just as (in a diverse world) there is more than just one notion of modernity, so is there more than just one possible future. Leibniz's concept of "possible worlds" carries within it the idea that a reason-driven shaping of the world will receive no reproof from reality.[48]

ORE FROM SUMATRA

While Leibniz is working on his letter to Landgrave Ernst and Arnauld, a few blocks away the Zellerfeld mint is bustling with activity. Unlike the Clausthal mint, the Zellerfeld mint is still molding and stamping silver coins the old-fashioned way, with the blows of a hammer.[49] Leibniz laments how quickly the precious metal flows out of the duchy and how it is losing value thanks to counterfeiting and the steady arrival of the silver fleets from the Spanish colonies. He reflects on the silver economy in the Harz not only in terms of its local significance, but also in the context of worldwide economic relations. With an alert eye, he follows the rapidly growing interconnection of the economy and the formation of a global market and sees risks and opportunities in equal measure.

In 1680 he had learned that the Dutch East India Company (Vereenigde Oostindische Compagnie, VOC) was planning to mine gold and silver ore on a large scale on the island of Sumatra (in present-day Indonesia). Samples of the ore had earlier ended

up in Clausthal for analysis by way of the Japan-born Dutchman
Pieter Hartsinck (Hartzing), who lived in the Harz and worked as
a mining adviser. Knowing all this, Leibniz advised Duke Ernest
Augustus to enter into a contract with the VOC. This would make it
possible for the ore coming from Sumatra to be brought to the Harz
for smelting. In exchange for the gold and silver thus obtained, the
Netherlands would receive shipments of canvas.[50] But the Dutch
mining expedition to Sumatra turned out to be a fiasco, and the ore
that ended up in the Harz, as Leibniz would learn, proved to be of
inferior quality.[51]

The Sumatra project is no exception. Leibniz never looks at
things in isolation; he always sees the far-reaching ties that stretch
around the globe. In order to counter the glut of cheap silver from
America, he petitioned Emperor Leopold in Vienna with the idea
of unifying all the German mining regions.[52] Later he will propose reactivating the old Hanseatic commercial network in order
to establish a direct trade route to China by way of Russia. This
would make the German territories independent from the Atlantic
powers, who control the sea route around Africa. In this way, too,
the Holy Roman Empire could profit from the Spanish silver mined
in America, which is regularly transported via Europe to China.
Leibniz is on the lookout for ways to ensure that Central Europe,
the "hinterland" of globalization, is connected to this worldwide
network. In the coming years and decades, the global players Russia
and China will take on an ever more important role in his plans.

And the Spanish ships in Veracruz? While Leibniz in Zellerfeld immerses himself in his worlds of words and paper, they're still
sitting in the harbor, waiting to set sail from Mexico and cross the
Atlantic, bound for Europe.

CHAPTER 3

HANOVER, AUGUST 13, 1696

THE SLEEPING WORLD, OR EVERYTHING IS FULL OF LIFE

As flies to wanton boys are we to th' Gods
They kill us for their sport.

SHAKESPEARE, *KING LEAR*, ACT IV, SCENE 1

BURNED OUT AND OVERHEATED

"Started a diary today," read the words written in dark brown ink on a sheet of folio paper, "in order to keep an accounting of the time I still have left."[1] Of the time I have left—Leibniz writes these words because he believes he could die soon. His potential demise no longer seems all that far off. Today is August 13 according to the Gregorian calendar, or August 3 according to the Julian calendar still in use in Hanover.[2] A few weeks earlier, Leibniz turned fifty. He feels weak, worn out, and empty. He can barely work, is jittery and distracted. Even math has become a torment for him. For the first time in his life, he can no longer rely on his energy and powers of concentration. He has felt unwell for far too long. Now it's time to put an end to it. Leibniz decides to keep a diary as a way of writing himself out of the crisis. And this Monday is when the diary begins.

Gottfried Wilhelm Leibniz's first diary entry,
dated August 3/13, 1696.

How Leibniz spends this day, and what he writes about it, is symp-
tomatic of his situation and of the way he leads his life. Exhausted
by too much work, he means to pull himself out of the rut he's in
through even more work. To fight fire with fire, one might say. Thus
we encounter a hyperactive Leibniz who increases his activity even
further by writing about it in the form of a diary. And in what he
undertakes on this thirteenth of August, we can see a reflection of
his life crisis. On this summer day, subjects like life and death, body
and soul, and perception and consciousness all play important roles.
What Leibniz thinks about and discusses with others today doesn't
just have abstract, theoretical significance for him. He presses far
ahead with his thoughts. Just how creative his crisis will ultimately
turn out to be is evident in the fact that he begins to develop ideas
that will one day become critical to the discovery of the unconscious.

It has been more than a year since Leibniz started consulting, by
letter, two doctors: Justus Schrader in Amsterdam and Conrad Barthold Behrens in Hildesheim. To them, he has complained of persistent hot flashes and constipation, a bitter ink taste in his mouth,
and a mounting disinclination toward work. Per remote diagnosis,
he was told he had a case of *affectus hypochondriacus*.[3] So Leibniz was
only imagining that he was sick? Not at all. In the seventeenth century, hypochondria was understood as a complex of ailments that
included flatulence, belching, heart trouble, and insomnia, combined with psychological symptoms including persistent anxiety. In

its advanced form, it resembled what today we would call a severe case of depression, which at the time was referred to as melancholy. Leibniz attributed his desolate condition primarily to the sedentary lifestyle (*vita sedentaria* or *vie sédentaire*) necessitated by his intellectual work.[4]

Behrens and Schrader suggested remedies that were meant to bring the relationship between acidic and alkaline forces back into balance. They recommended a diet of goat's milk and an herbal bath with animal tooth powder; bloodletting was likewise considered.[5] Whether Leibniz followed this advice, the record doesn't say. The idea of drinking a concoction of wormwood, mint, white wine, and cinnamon syrup early every morning, followed by a vigorous half-hour walk, certainly can't have been to the liking of the chronic late sleeper. In any case, as of this Monday in August, his condition has shown no substantial improvement.

Death is a part of Leibniz's daily life. In the seventeenth century, the mortality rate is higher in every age cohort than it is today. It's not uncommon for young people to die suddenly. Daily living conditions are hard, and in this crisis-plagued century, disease, war, and the Little Ice Age only make things worse. Leibniz has already seen many companions fall by the wayside: Huygens, Arnauld, and Landgrave Ernst, to name just a few from recent years. And his current lord and protector, Ernest Augustus of Brunswick-Lüneburg, who was recently granted the title of Elector of Hanover, is gravely ill. Often one doesn't know if someone is even still alive. It doesn't take long for a man to be considered dead. Leibniz has heard from Thomas Burnett of Kemney that a rumor of his own death has been going around in England.[6] This rumor, of course, has been greatly exaggerated. By how much, Leibniz doesn't know, but he intends to make use of the time he has left. But how to get out of this permanent state of crisis? It's a well-known fact that a subconscious fear of grave illness or death can lead to heightened self-awareness. Leibniz is determined to put his life back on a firm footing by managing his time in a deliberate way. Keeping a diary seems to be the solution.

IN THRALL TO THE COURT

Leibniz writes on the evening of August 13 that "this morning" he received news of his promotion to the position of Geheimer Justizrat.[7] It's no coincidence, then, that he begins keeping his diary on this particular day. Finally a promotion, finally more pay. He has had to wait a long time for this. After the failure of his experiments in the Harz, the task of writing a history of the Welf dynasty moved to the forefront of his official duties. While on an extended trip that took him to southern Germany, Austria, and Italy between November 1687 and June 1690, he was able to visit libraries and archives and gather ample source materials for the project. Since then he has been keeping up a brisk correspondence with genealogists and librarians all over Europe. This morning, for example, he receives a letter from Antonio Magliabechi, head of the grand ducal library in Florence.[8] Magliabechi fills him in on the latest news from the world of Italian scholarship. Over the course of his long trip, Leibniz managed to prove, thanks to his study of genealogical sources, that the North German Welfs and the Italian House of Este had shared ancestry. After Leibniz published his findings, Magliabechi played an active role in helping to disseminate them in Italy. This not only garnered Leibniz much acclaim in the scholarly world but also helped to arrange a marriage between Rinaldo III of Modena and the Hanoverian Princess Charlotte Felicitas. The union no doubt played a decisive role in securing Leibniz his promotion.

Leibniz seems to have finally arrived, so to speak, at the court in Hanover. He lives in the Leine Palace, which gives him access to the court library and puts him in the immediate vicinity of his prince. Nevertheless, he is still kept at arm's length. He gets yet another taste of such treatment on this August 13. News of the promotion, approved back on July 17, is conveyed to him through official channels, following strict protocol: he receives word from Geheimer Kammersekretär Jobst Christoph Reiche, acting on orders from Franz Ernst, Reichsfreiherr von Platen, who serves as chairman of the privy council, Oberhofsmarschall, and Premierminister, and who in turn is acting on orders from the Elector of Hanover. Despite his rise from Hofrat to Geheimer Justizrat, Leibniz continues to occupy only a middle position in the court hierarchy.[9] He often feels misunderstood among the court officials. It distresses him, he will write to Burnett later this year, not to live in a big city like Paris or London, where he would have ample opportunities to engage in discussion and trade ideas with like-minded men of learning. Here in Hanover, by contrast, he finds hardly any suitable conversation partners, with the exception of Electress Sophie. One ceases to be considered a courtly gentleman as soon as one starts talking about scientific matters.[10]

Not that court life is by any means alien to Leibniz. To the contrary, as a purveyor of ideas representing the elector's noble person he is eagerly sought after whenever a speech is needed for a wedding or funeral in Hanover; or if an inscription is needed for a commemorative medal, or a motto or slogan for a theatre building; or a cultural program has to be put together to accompany a masked ball.[11] Leibniz is likewise an expert in court diplomacy. Just a few weeks earlier, looking for a means to derive special privileges for the elector's envoys, he brought out a book on diplomatic customs at the Roman Curia.[12] On the whole, though, Leibniz's elevation to Geheimer Justizrat on August 13 means primarily one thing: an even heavier official workload. It will inevitably carry with it— as Leibniz is all too aware—an increase in his already numerous duties: as jurist and legal expert, as aide and adviser, as librarian and court historian. A secretary to power is never off the clock. He who would influence the throne must never rest. As welcome as the promotion is, indeed as long desired, at the same time it adds to the burden of obligations and commitments that the rolling stone must bear.

THE FREEDOM OF VOLUNTARY COMMITMENT

But that's not all. His self-imposed duties as a scholar far surpass his range of activities as court official. Despite his growing burden of *occupationes*, as he'll write to a friend the following year, he continues to grant himself "much freedom."[13] Science is his "foible"—this he stresses again and again. Nothing means more to him than the freedom of intellectual inquiry, but this also means a steady load of extra work. He wouldn't dream of resting on his laurels. By this point, Leibniz is a mathematician esteemed throughout Europe—and in England, among the followers of Newton, he is increasingly feared. His method of differential and integral calculus has by this point become successfully established in several places. But fame brings obligation. Again and again he is asked to take a position on some issue or another. The mathematician Angelo Marchetti from Pisa, who has just brought out a book on differential mathematics, is hoping for a favorable response from Leibniz. Marchetti's letter reaches him on the morning of August 13, enclosed with the letter from Magliabechi.[14] Leibniz has already heard about the book from other mathematicians and immediately expresses doubts about Marchetti's approach.[15] He seizes every opportunity to test how his method holds up. Just recently, in the June edition of *Acta eruditorum*, Johann Bernoulli had posed the so-called brachistochrone problem—how to calculate the route that a body acted upon by gravity must take in order to move between two given points in the shortest amount of time. A mere week later Leibniz was able to solve the problem by providing a differential equation for the curve in question.[16]

Full of zeal, Leibniz rushes headlong into every mathematical competition. But this eagerness is getting to be a problem. Soon he'll hardly be able to keep up—on the one hand, he continues to test out more new fields of application for his calculus, and on the other, he has to keep pace with the growing number of inventive

young mathematicians. Ambitious colleagues like Jacob and Johann
Bernoulli and Guillaume François Antoine de L'Hôpital are moving at breakneck speed, building on Leibniz's calculus and becoming rivals, edging in on his own turf. On top of this, Leibniz, with
his *scientia infiniti* and his *analysis situs*, continues to hold out the
prospect that he might achieve major breakthroughs in other areas
of mathematics as well. By this point, however, he realizes that he
has overextended himself. As he writes to Augustinus Vagetius on
June 15, his *mathesis universalis*, once intended to be a comprehensive work, will now be confined to a treatment of purely quantitative
values of algebra and arithmetic.[17]

As ever, mathematics is just one subject of interest to him. Across
almost the full breadth of his universal intellectual horizon, Leibniz
promises to invent or discover new and revolutionary things. That
burden increases alongside admonitions from his friends to follow
through on things he'd promised long ago. Unanswered letters pile
up on his desk; it's no rarity for there to be more than thirty of
them.[18] On this August 13, the pile is again massive. Leibniz is
plagued by his own productivity, constantly torn between unfinished, stalled-out projects and a wealth of new inspirations and
ideas. The oscillation between his two main areas of activity—the
elector's court and the republic of letters—has become a punish-
ing tug-of-war that takes an increasing toll on his health. He finds
himself caught in a vicious cycle: the growing work pressure saps his
strength, and out of fear that it will give out completely, he hastens
to complete what he has begun—which increases the pressure and
weakens him further.

The connection between his overworked, overwhelmed condition and his perilous health has not escaped him. He did, after all,
declare to his friend Thomas Burnett of Kemney in March that he
would see the plans he had already sketched out through to fruition
if only death would give him time. In exchange, he would promise death not to begin anything new, even if the reprieve this deal
granted him were a long one.[19] Since then, however, Leibniz has

repeatedly broken his pact with the Reaper. But there's method to his madness. Instead of listening to his doctors, he prefers to come up with his own course of treatment—and a very special one at that.

WATERWORKS AT HERRENHAUSEN

Leibniz's approach to self-therapy is as simple as it is serious. He plunges ever deeper into the vicious cycle in order to break out of it. Fighting overwork with even more work is a thoroughly common course of action (as much so today as back then); many who find themselves stuck in the muck choose to try to lift themselves out singlehandedly. Often one's relationship to time plays a crucial role, as is the case with Leibniz. His diary is meant to help him use his time effectively. *Pars vitae, quoties perditur hora, perit* (A portion of one's life is lost when an hour is wasted)—not for nothing will Johann Georg Eckhart have these words set on Leibniz's coffin.[20] And so Leibniz, as soon as he has read the letters from Magliabechi and Marchetti, immediately sets out to see Kammerpräsident Friedrich Wilhelm von Schlitz genannt von Görtz, one of the highest-ranking officials in Hanover, in order to speak with him about an important matter. It concerns the improvement of the waterworks in the gardens around Herrenhausen Palace, the summer residence of the elector and his family.

Following the model of the gardens at Versailles and at the Villa d'Este in Rome, the Herrenhausen garden will provide the setting for court celebrations and amusements, and the plans for its layout are suitably grandiose. A large waterfall, a grotto, and a garden theater have already been built, and since 1694 work has been ongoing on a spacious gallery that is meant to have the character of a pleasure palace (*maison de plaisance*). Ingenious water features—streams, springs, and fountains—are all meant to advance the Hanoverian court's ambition to be counted among the most magnificent in all of Germany. Leibniz was grateful to receive the job of contributing

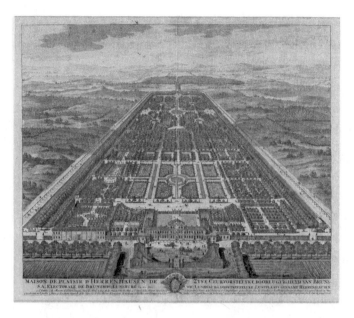

*Idealized rendering of Herrenhausen Palace and
surrounding gardens, c. 1708.*

to the plans from his lord and employer, Elector Ernest Augustus. His primary task is to help design a system that will channel water into the gardens with sufficiently high pressure to power
the fountains and *Springwerke*. Two specialists in hydraulics have
just arrived from the Netherlands: Bernhard Schotanus van Steringa from Leeuwarden and Balthasar van Poelwyck from Amsterdam. First Leibniz discusses the Dutch experts' proposals with the
Kammerpräsident. After that he travels out to Herrenhausen with
the two guests in order to survey the terrain; soon enough the Kammerpräsident joins them. Schotanus and Poelwyck argue for the use
of windmills. Their proposal is to use wind power to draw water
from the Leine into a trench and divert it to the palace; there the
water would be drawn up into an elevated basin by a horse-powered
pump system, which would ensure that it would flow from a great
enough height to power the pumps for the fountains.[21]

Leibniz objects that there's hardly enough wind to make this proposal practical and that the use of horses is too costly. Instead he

offers up his own idea of using a canal to divert the water to Herrenhausen and onto the waterwheel rigged to power the pumps in
the garden. A few days earlier he had been advocating for the construction of a canal from the Leine to the palace garden, in a meeting with Generalfeldzeugmeister Andreas Du Mont. Traffic on the
Leine would be undisturbed, and in addition the water, before being
directed back into the river, could flow into other areas of the gardens via additional canals that could serve as transportation routes
or for gondola rides and festive nighttime light displays.[22] Yet again
Leibniz attempts to unite various interests, to solve an engineering
problem and at the same time to cater to the Venice-mad court society's desire for amusement. As with the Harz project, the intention
is also to find a practical way to improve the world, even if it's only
the highly confined world of diversions crafted for the sake of a
Baroque display of power. It's astounding, all the things that Leibniz takes on this Monday. He's full of drive, there's no holding him
back—and the morning isn't even over yet.

SUMMERTIME CHATS

Around noon, still in Herrenhausen, he meets with the mistress
of the palace, Electress Sophie of Hanover. The consort of Ernest
Augustus holds Leibniz, who is sixteen years her junior, in high
esteem. Over the years a close friendship has formed between the
polymath and the noble lady. Whenever an opportunity presents
itself, the two of them meet to talk about politics, news from around
the world, and the latest court gossip; they even discuss science and
philosophy. Their correspondence began when Leibniz was traveling in southern Germany, Austria, and Italy, and since his return
they have regularly exchanged letters when they're not in the same
place. Leibniz admires the electress's keen intellect; Sophie takes
enjoyment from Leibniz's learned presentations and also appreciates
that he not only possesses a masterly command of courtly manners

but also, like her, knows how to defy them at times with ample irony and mockery. Occasionally the two of them go for strolls together on the extensive garden paths at Herrenhausen, or they meet, together with others, for genial social gatherings. Four years earlier Leibniz is said to have challenged a guest, Carl August von Alvensleben, to find two identical leaves in the garden. The guest was unable to do so, which for Leibniz served as an example of the unmistakable individuality of all living things.[23]

What's on the docket today, however, isn't mere chitchat in the garden. Sophie receives Leibniz in her newly built, richly decorated "mirror cabinet" (*Spiegelkabinett*), possibly occupied—as she so often is—with a piece of embroidery. At Herrenhausen, the usual court ceremony is relaxed; inside the electress's apartments, niches open up for the free exchange of views. Here one need not fear the censure of church or state; in the inner chambers of power, these constraints can be suspended, at least temporarily. Today ideas are floated that run counter to the prevailing views of the Christian church, whether Catholic or Protestant. They come from a letter that Sophie gives Leibniz to read. It was written by Sophie's niece in France, Duchess Elisabeth Charlotte of Orléans.

The author of the letter, who is also known as Liselotte von der Pfalz, has been married to Duke Philippe of Orléans, the brother of Louis XIV, since 1671; most of the year she lives in the palace at Versailles. She is known to be free-spirited and isn't afraid to speak her mind. In her letters to her aunt, she often vents her criticisms of the deceit and dissembling among the favorites and courtiers in the shadow of the Sun King. Estranged from her husband, condemned to aimless perpetual pleasure in the golden cage of the court, she frequently spends her time on extended hunting excursions. On carriage rides, she occasionally whiles the hours away with stimulating literature. Elisabeth Charlotte of Orléans, eminently unhappy, seeks consolation in reading and reflection. This summer she is re-reading *The Consolation of Philosophy* (*Consolatio philosophiae*) by the late Roman philosopher Anicius Manlius Severinus Boethius; more specifically,

she is reading the German edition of 1667, which Franciscus Mercurius van Helmont, one of the translators, had given her twenty-five
years earlier.[24] Along with the Boethius translation, she has also been
perusing other writings of Helmont's.

Helmont is an eclectic scholar and theologian, a free spirit unat-
tached to any church who has also studied mysticism and esotericism
and is well versed in hermeticism, theosophy, and the Kabbala. The
intellectual maverick, born in 1614 in Vilvoorde in Flanders, is by this
point a professed Quaker, while on the side he also occupies himself
with alchemy. At the center of his theosophist philosophy is the doctrine of reincarnation, which teaches that a person's soul passes into
another body after death. Leibniz first met Helmont, persecuted by
the church and on the run from the Inquisition, in Mainz in 1671.
Like Leibniz, Helmont is a rolling stone, restlessly shuttling between
scientific circles and the courts of Europe's rulers. Helmont's philosophical and mystical writings, from a treatise on hell to a book on
creation and the apocalypse (*Seder Olam*), could, to quote Leibniz's
assessment, "make a normal theologian fly into a rage."[25] Electress
Sophie is also familiar with them. As her "literary sampler,"[26] Leibniz
has read and reported on Helmont's work for her. Sophie, who values
challenging thinkers, invited Helmont to Hanover back in March
for shared discussions with herself and Leibniz. Many conversations
took place between the three of them, often beginning, to Leibniz's
displeasure, as early as nine o'clock in the morning.[27]

When Electress Sophie finds out that her niece is studying Helmont's teachings, she asks her to write and tell her what she thinks
of them. Both noble ladies are fairly skeptical of the independent-
minded scholar's speculations, but they respect his positive demeanor
as well as his demonstrative air of relaxed contentedness, and they
would be all too glad to learn the secret behind this profoundly calm
peace of mind. When Elisabeth Charlotte's letter containing her
views on Helmont arrives in Herrenhausen, the aged eccentric himself happens to be there as well. At eighty-two, he is still in robust
health, doesn't shy away from long journeys, and is in Hanover for

the second time this year. Helmont, however, is not present when Sophie hands Leibniz her niece's letter. She'd like Leibniz to have a chance to form his own thoughts about the letter from France before she brings Helmont into the discussion.

Leibniz reads the letter in the presence of the electress and immediately copies out an excerpt for himself. "I can't really wrap my head around what Mons. Helmont is saying," the excerpt from Elisabeth Charlotte's letter begins, "because I cannot understand what [a] soul is and how it can go into another body."[28] The duchess can't piece together how Helmont's notion of metempsychosis is supposed to work. Instead she believes that after death we dissolve back into those elements from which we came, and that what remains of us serves as nourishment for the plants or other creatures. Not nature but rather God's grace alone can make the soul immortal. Nor does Elisabeth Charlotte consider it possible to derive a sense of God's justice from human notions. She's also not clear on the idea of original sin: why was the blind man punished with blindness from birth, even though neither he nor his father had sinned (John 9:2)?[29] Sophie's niece at the court of Versailles isn't just questioning the notion of reincarnation, which according to the official view of the church is deviant; she is also, with her criticism of the doctrine of original sin, questioning central tenets of the faith of that same church.

For Leibniz, Helmont's claims and the duchess's commentary are a welcome invitation to expound upon his own views. In an initial statement, which he prepares for the electress the following day, he responds at length to Elisabeth Charlotte's objection that God's actions or inaction cannot be judged by human standards. Not only are there truths, for example in mathematics, that are equally eternal and unchanging for man and God both; it is also the case that man, owing to his limited reason, cannot fully comprehend things. If we had God's all-encompassing perspective, then everywhere we looked, where we now see only chaos and disorder, we would behold justice, order, and beauty.[30] It seems easy enough for Leibniz to incorporate both the teachings of Helmont and reservations about them into his own philosophical system of a world of divinely guaranteed, harmonious order. But what about reincarnation?

THE REAWAKENING OF THE FLIES

As the discussion of this topic continues into the afternoon, questions about life, death, and dying become the focal point. Leibniz can understand the duchess's assertion that the human soul seems to be mortal without the intervention of divine grace. Yet he responds that on closer examination, one can see that "the destruction of the soul does not follow from the scattering of the parts of our body."[31] A year earlier Leibniz had expressed his views on this subject at length in the treatise *Système nouveau de la nature et de la communication des substances* (*New System of the Nature and Communication of Substances*).[32] *The Discourse on Metaphysics* composed in the Harz remained unpublished; eventually Arnauld had stopped responding to Leibniz's letters. Now, however, with the publication of the *New System*, Leibniz's metaphysics is available to the public for the first time and in more developed form. The exchange of views with Sophie, Elisabeth Charlotte, and Helmont gives him yet another opportunity to shape and hone his arguments.

Like Helmont, Leibniz believes in the immortality of the soul, but unlike him, he does not believe that the soul transmigrates to another body. In death, Leibniz says, body and soul stay together. In his view, death does not signify the death of an individual but merely a minimization of its size, a reduction of its parts to an infinitely small degree. It continues to exist but is invisible to human senses, just as it was before birth. Accordingly, birth and death are not beginning or end points; they are merely moments of transition, in which the living thing either blossoms and grows or, alternatively, shrinks back into itself and becomes smaller. Once created by God, no individual can be destroyed except by him. Unless this happens, the individual will not perish but will pass through phases of transformation—just as a snake sheds its skin several times, a caterpillar turns into a butterfly, or a larva into a fly.

In this, Leibniz sees further proof that nature makes no leaps. Death is usually too intense for one to be able to discern in detail the sudden shrinking, as one does with slow growth, but the same fundamental principle applies: the uniformity of all natural processes. Thus it's hardly possible to determine the exact time of death. Simple creatures often linger for extended periods of time on the threshold between visible and invisible movement. And at this very moment—as if out of nowhere—an old acquaintance reappears: in the *New System*, Leibniz offers up as an example the reanimation of flies doused in water, which begin to stir if one coats them in powdered chalk.[33] Leibniz is familiar with this phenomenon from his reading of scientific literature, where he has encountered several examples. Robert Hooke, whom he met in London in the spring of 1673, gives an account in his *Micrographia* (1667) of flies that become paralyzed in the cold or seem to drown in liquid but that later, when one warms them up or lets them dry out in the sun, come alive again.[34]

And so one might say that Leibniz doesn't save only the soul from its demise; he also saves the living thing that goes with it. And preserving the soul does not, contrary to what Elisabeth Charlotte suggests in her letter, require an extraordinary act of divine grace. Everything happens in a natural manner. Even if, in Leibniz's view, there is no direct connection between soul and body, their unity in a given living thing nevertheless persists beyond its death—of this he is convinced. A death that is no death. This beautiful paradox holds meaning for Leibniz that is more than just philosophical or theoretical; it has very concrete significance for his own life as well. At a time when he is threatened by illness and death, he turns increasingly to metaphysical meditations on birth, life, and mortality. Reflected at the highest level of abstract argumentation is the ebbing of his health and capacity for work. In other words: at the pinnacle of his creative crisis, his philosophy of being and of life also reaches its pinnacle—for the time being. Yet again Leibniz has occasion to see that bad things can also have their good side.

THERE IS NOTHING DEAD IN THE WORLD

That same day Leibniz speaks with Helmont, lays out for him Elisabeth Charlotte's objections, and takes note of his responses. Three days later he meets with Sophie again in her apartments in Herrenhausen, relates Helmont's answers, and hands her the notes he made so that she can use them for her reply to the duchess. He keeps a copy for himself.[35] In this way a four-way conversation develops among Sophie, Elisabeth Charlotte, Leibniz, and Helmont that will carry on into the fall. Leibniz functions as a pivot, a conversational point of intersection, even if he sometimes finds it hard to follow Helmont's explanations. At times with Helmont, he writes, he has taken up his quill and "tried to draft, point by point," the older scholar's line of argument, as well as his own objections, "but in the end generally could not do so."[36] Tirelessly he relays questions and answers back and forth among the other three and inserts his own considerations as well. What emerges is a unique discourse that gives us deep insight into the intellectual life of the Baroque age on the cusp of the early Enlightenment.

Leibniz's arguments lead to the conclusion that just as people don't really die, animals don't either. Duchess Elisabeth Charlotte is extraordinarily happy with this thought; she immediately thinks of her beloved lapdogs, who provide welcome diversion during the endless hours of boredom at court. "It's a great consolation to me that animals don't die completely, thinking about my dear dogs."[37] "A great consolation to me"—this sounds like Boethius's motif of finding consolation in philosophy. For Leibniz, in any case, what follows from all this is that nothing in nature is really dead. He sees support for this theory above all in the studies in microscopy with

The blue fly. Robert Hooke, Micrographia: Or Some Physiological Descriptions of Minute Bodies Made by Magnifying Glasses . . . *(London, 1667).*

which he is so well acquainted. Thanks to Hooke, Swammerdam, and Leeuwenhoek, who many years ago invited him to peer into their microscopes, he has delved into heretofore unknown microscopic worlds and has seen how lice and fleas can transform into giant hairy monsters, and how even the harmless housefly can come to resemble a spiny winged beast.

In Leibniz's view, these instruments also yield information about how each individual life develops. What they make it possible to see strengthens him in the assumption that every living being is already prefigured in miniature form in the female's egg or the male's sperm, and that upon birth it merely has to unfold or develop outward. He then simply reverses this thought about preformation and applies it, as described above, to death, which for him is nothing other than the starting point of a process of *inward* development or involution. The only difference from the process of outward development is that the involution cannot be observed. In the microscope's enlarged visual field, a water droplet becomes an ocean teeming with tiny creatures—Swammerdam is one of the first to describe what will later be called bacteria. Even a grain of sand seems to harbor an entire world within itself. No portion of matter is so small that it can't offer room to a multitude of living creatures. Leibniz takes this line of thinking even further: every one of these worlds carries within itself additional worlds containing countless creatures, ad infinitum. For Leibniz, nature ultimately consists of an infinite number of worlds nested inside one another, even if, because sense perception is ultimately limited, this infinity cannot be observed under the microscope.[38] The nesting continues with the organs: every organism carries an endless number within itself, such that there is practically no end to living matter and only parts of an organism can ever be destroyed, never the organism itself.

SCREAMING MACHINES

So far, so good. If animals, like people, ultimately do not die and the human soul is immortal, then do animals have souls too? It's well known that Descartes and his followers' answer to this question is a resounding no. To the Cartesians, animals are mere machines, "artificial clockwork mechanisms," as Leibniz expounds, "driven by fire and wind," "without any sensation," so that when they cry out, they "feel no more than an organ pipe."[39] At its very foundation, Leibniz's natural philosophy is so constituted as to firmly contest this claim. And in this respect, he knows himself to be in agreement not just with Helmont but with Sophie and Elisabeth Charlotte as well. According to Leibniz, people are not unique in having souls; for him, souls are present everywhere in nature, though they differ in complexity. What Leibniz consistently describes to his noble auditors as a soul corresponds in his metaphysics to the simple, indivisible substances, the "animated" points that each in its own way mirrors the whole universe within itself and possesses an intrinsic energy. This is what distinguishes living beings from dead matter. By this logic, if we follow Leibniz's metaphysics as applied to the natural world, plants and animals also possess sensation and thus a kind of soul, at least in a simple form.

In this connection, Leibniz makes use of the concept of perception, which for him is central. Perception signifies more than mere sensory awareness or feeling; rather, it describes the condition in which a simple substance or soul is cognizant of and expresses its entire surroundings. Inherent in this condition is a *telos*, a goal-directed element, namely a striving (which Leibniz calls "appetition") to move from one state to the next, and the constant restlessness that goes with it. In this respect, humans, animals, and plants are not fundamentally different from one another; the difference is that the latter two's perceptions are less distinctly pronounced than are those of humans—Leibniz speaks of "little perceptions" (*petites perceptions*). Thus the perceptive faculties of different life-forms vary only in degree. If, however, the living being reflects on the acts of perception, that is, observes them on a second level (what Leibniz calls "apperception"), then consciousness and memory can emerge. Therein lies the actual difference for Leibniz between humans on the one hand and plants and animals on the other. All living creatures occupy a position on a hierarchical ladder of increasing perceptive capability, which extends from the confused perception of lower life-forms all the way up to the self-reflexive consciousness of humans. The ladder doesn't stop with humans, however; above them are the angels, whose cognitive ability far exceeds human reason. But even on this rung, the cognitive faculty is not yet fully developed. Only within God's infinite mind does reason shine in all its brilliance; only here have wisdom and knowledge reached absolute perfection.

AUTOBIOGRAPHICAL FLASHBACKS

If self-reflexive consciousness is present, then past and present conditions can be brought together and tied to future expectations. That fulfills an important precondition for the constitution of the

Johann David Schubert,
"Leibniz Chooses between the
Old and New Philosophy,"
copperplate engraving,
published in Karl Gottlieb
Hofmann, ed., Pantheon der
Deutschen, 2. Teil *(Chemnitz,*
1795), unpaginated.

individual self, expressed in the self-characterization *I* and organized through acts of remembering. We see a good example in
Leibniz's own behavior during the crisis years of 1695–96. It's a
time of intense reflection on the past. Accordingly, most of his
autobiographical musings come from these years. In letters to
friends and fellow scholars, he looks back again and again at his
childhood and youth. The same tropes and images keep popping
up: the highly gifted young autodidact in his father's library, or the
precocious writer of poems, hundreds of verses long, in Latin. One
scene in particular stands out: the fifteen-year-old Leibniz wanders
through the Rosenthal Wood, outside Leipzig. Inside he is torn,
desperately struggling to decide between the old philosophy and
the new. Aristotle or Descartes? Scholastic or mechanistic natural
science?

Logic and conceptualism, or observation and experiment? Leibniz chooses the new teachings without, however, discarding the old;
rather, he intends to connect the two. In the crisis period around

1696, he deftly dramatizes the years-long development of his own philosophy by projecting his memory back in time and boiling things down to a central decisive moment.

The philosopher at the crossroads: in the face of poor health and an ongoing creative crisis, Leibniz seeks self-assurance by examining his own life to that point. Who knows what the future holds? A latent fear of death contributes to his impulse to craft his own image for posterity. And he will do it with success. Leibniz's account will gratefully be taken up in later years, as evidenced by a copperplate engraving, made in 1795 and serially reproduced, that depicts the famous philosopher at the moment of his fateful decision, reclining under a tree in the Rosenthal Wood. At a time when one is often reflecting on birth, life, and death (as Leibniz is currently doing), the borders between what one has experienced and what one has read can certainly blur, whether consciously or unconsciously. Thus the episode mentioned in the *New System* concerning the flies that were brought back to life, which Leibniz is familiar with from his reading, becomes, in a letter to Electress Sophie, a personal experience: "When I was a little boy, it gave me joy to see how drowned flies were resuscitated when I dropped them in powdered chalk."[40] Maybe as a boy he actually took more joy in bringing flies back to life than in killing them. By this point, in any event, the fly has become intimately connected with Leibnizian metaphysics, playing a significant supporting role in the drama of eternal return, of perpetual becoming and passing away.

BLURRED INFINITIES

Remembering and forgetting are central in Helmont's theory of reincarnation as well. Not being able to remember having lived before was "poor consolation, for one retains only the knowledge that one will die but knows nothing of living again."[41]

Thus writes the duchess in the letter that Leibniz copies in the

"Comforting Wisdom,"
copperplate engraving used
as the frontispiece to Anicius
Manlius Severinus Boethius,
Consolatio Philosophiae oder
Christlich-vernufft-gemesser
Trost und Unterricht in
Widerwertigkeit und
Bestürzung über dem
vermeinten Wohl-oder
Übel-Stand der Bösen
und Frommen . . . , *trans.*
Franciscus Mercurius Helmont
(Lüneburg, 1697).

presence of the electress on August 13. In an earlier letter, Elisabeth Charlotte was even more forceful in her criticism: "It seems to me that it wouldn't be worth the trouble to live again so often only to die so often, since all one would get out of it is the trouble of living, but not the consolation of knowing that one doesn't completely perish, since, seeing as how one can't remember anything, it's just the same as if one hadn't ever been."[42]

Without an active memory of past lives, the whole thing seems to be pointless. Or to put it another way: without that active memory, Helmont's idea of reincarnation seems unable to dispel the noble ladies' melancholy and thus offer them any Boethian consolation. Leibniz doesn't see it that way. He answers the electress and the duchess as follows: "rather think of it like this, that everything that has ever happened to us remains imprinted on the soul forever, even if we don't always recall it; just like we know a lot of things that we don't remember if we aren't properly reminded of them, or unless a particular cause makes us think of them."[43] According to Leibniz,

memories don't disappear with death but instead remain imperceptibly present in the soul. Our memory, says Leibniz, is capable of
storing more than we can actively recall.

Memories can lie dormant, unnoticed, within us. Many things
we perceive unconsciously. Here Leibniz draws an explicit parallel
to the diffuse, small perceptions of animals, plants, and microbes.
When asleep, unconscious, or in a daze, the human individual finds
themselves in a certain respect to be on the level of a simple life-
form. Thus the whole world is by varying degrees asleep and awakes
from this slumber gradually, in accordance with the tiered order of
life—Leibniz likely took the term *étourdissement* (dizzy spell or daze)
directly from Helmont.[44] Only if the world were to rise to the level
of God and attain the condition of divine perfection (which, however, it will never do)—only then would it completely arrive at clear
understanding and alert consciousness. In Leibniz's conception, the
small, unnoticeable perceptions occur in uninterrupted succession.
That means they don't stop when we faint or are overcome by deep
sleep. While we are dreaming, notions can seep into our consciousness that we perceived indirectly or that found their way into our
memory unnoticed on some long-ago occasion. Indeed, the dream
can even amplify them, and all of a sudden something steps into the
foreground of our perception that we were never aware of before.

Leibniz will later seek to illustrate the idea with the famous
image of the roar of the sea (*bruit de la mer*): one could not hear the
crashing of a large wave if the many small waves that make it up did
not each produce a tiny sound inaudible to human ears.[45] Hearing
is yet another thing that Leibniz will later concern himself with, in
his experiments in acoustics and how sound is generated in the ear.
Moreover, he is convinced that our confused perceptions are the
result of impressions that arrive from throughout the entire universe
and have an effect on us, even if, like waves, they grow weaker the
farther they have to travel to get to us.

In reflecting on these matters, Leibniz will have a decisive
influence on the emergence of psychoanalysis, in particular the

interpretation of dreams. One of its leading exponents, Ludwig Binswanger, will refer to the influence that remote occurrences seem to
have on dreams, which in his view capture the events of the cosmos
in wave form. "Thus man appears embedded in 'cosmic' activity,"
he will write, "which, in waking, he takes less notice of than while
asleep, due to the stronger stimuli in his immediate surroundings.
Here man is placed in the universe, I'm tempted to say, like a living,
animated seismograph, a notion which later on we find interpreted
in more or less spiritual-mystical or biological-magical fashion particularly in Plotinus, in the Renaissance philosopher Campanella,
in Schelling, Carus, G. H. Schubert, Novalis, in Fechner and in
Schopenhauer, whereas Leibniz has given it the most logical and
profound, *purely* spiritual-metaphysical expression."[46]

SOUL-DESERTS

As we've seen, the consolation that the duchess and the electress
take from the discussions at Herrenhausen seems to have been distinctly limited. Again and again Helmont is the victim of their
biting ridicule, although at bottom they envy the always even-
tempered and contented old man. Leibniz often has trouble following his arguments; some parts he finds stimulating, while others
seem nothing more than abstruse prattle. But he too admires the
fact that Helmont has no fear of death. And their summer dia-
logues are certainly fruitful. Together with her daughter Sophie
Charlotte, now Electress of Brandenburg, Sophie is seeing to the
publication of a new edition of Helmont's Boethius translation.
Leibniz contributed the foreword, which he wrote in Hanover back
in June. In it he mentions a person who was cured of a "Melan-
choley" with the help of the little volume—and he may well have
been referring to himself.[47]

Crises can give rise to new things and bring others to a close. This
summer turns out to be a crucial maturation period for Leibniz's

metaphysics. The diary does in fact help him overcome the exhaustion and discomfiture of the last months and years. Gradually he seems to succeed in finding a satisfying answer to the fundamental questions of metaphysics. The various aspects and perspectives that he has wrestled with for so long seem now to come together to form a coherent image: Leibniz at first turns his back on Aristotle, but finds in Descartes and his ilk no satisfying answer to the question of what holds the world together in its innermost core. The world must be more than a mere aggregate, a heap of particles or atoms. But what actually are its smallest units?

By going back to Aristotle, Leibniz finds the answer in the proposition that substantial entities must contain within themselves some form-constructing element, an active principle: simple, indivisible substances (*substances simples*). These simple substances form that thing in the body which is called the soul. The organic, which in and of itself is infinite, is brought together in the soul to form a unity. This unity possesses a striving impulse to go from one state to another, and energy to effect this change. For the individual, the world is made up of a constant stream of perceptions of varying degrees of clarity. Every soul or simple substance represents the whole universe, and each regards the world from its own distinct point of view. This point of view, in its relation to the world as a whole, is like the point at the center of a circle, which unites within itself all the lines that one can draw to it from the circumference. In this way it extends in all directions, even though, as a mathematical (or more precisely, a geometric) point, it has no dimensions, neither height nor width nor breadth. In the field of term or propositional logic, such a viewpoint—or standpoint— corresponds to a subject that contains all its predicates within itself, a complete notion (*notio completa*). Biologically speaking, this means that, as previously described, every individual is subjected to a never-ending process of outward development and involution, which from a theological standpoint corresponds to the principle of immortality.

Thus Leibniz attempts to shine light on the subject from several angles—from ontology and biology, to mathematics and logic, and finally on to epistemology and theology—in order to make good on a claim of offering a universal explanation, for such is his ambitious goal. What has been missing up to this point, however, is a fitting term, one capable of gathering all this together. But then, in the middle of his existential crisis, the tide seems to turn. A little more than a year has passed since Leibniz first used the Greek-derived term *monas* to express, in a metaphysical sense, the unity and indivisibility of a simple individual substance. The term seems to him more functional and illustrative than *simple substance* or *substantial atom* because *monas* best characterizes the idea of a single unit that in a sense encompasses the plurality of things but cannot itself be broken down into smaller units. To put it another way, the unit comprehends the plurality without being a part of it.[48]

Later *monas* will become *monad* and will go on to become the core concept of Leibnizian metaphysics, which thenceforth, when dealing with the theory of substances in a more narrow sense, will be called monadology. And with that, a concept is put forth into the world that is hard to pin down and still poses riddles today. It's possible the polymath was trying to achieve too much at once, namely "to gather together in a single term the permanence of the ultimate, irreducible bearers of physical forces, the continuity of the bearers of life, and the unity of consciousness, and in so doing to prove their immortality."[49] Leibniz seemed himself to balk at the weight of meaning that this concept was meant to convey. For in the very spot where the word *monas* first issued from his quill, in a letter to Guillaume François Antoine de L'Hôpital dated July 22, 1695, the writing breaks off—almost as if Leibniz hesitated and didn't want to use the newly created term after all.[50] This wavering was no doubt also bound up with the lack of courage and self-confidence he felt in this phase of despondency and insecurity. No sooner had he set the term *monas* down on paper than he immediately put it back on the shelf. There it remains; the indecisiveness persists on this August 13.

HOW MANY LEIBNIZES IN ONE PERSON?

As afternoon sets in at Herrenhausen, the heat of the day increases, but the summer's peak has passed. All in all, it's been a rather cool season. It rained a lot throughout almost all of June, and only since July have there been extended periods of warm weather.[51] It's quite quiet in the gardens; fruit ripens and vegetables grow in the ornamental beds. Birds, bees, and butterflies find plenty to eat, while the hedges shelter myriad insects. Fish splash about in the ponds, and every now and then the buzzing of bumblebees can be heard. It's getting to be time to leave. The diary indicates that Leibniz, Helmont, and the Dutch hydraulic and windmill experts Schotanus and Poelwyk leave the summer palace grounds in the afternoon and set out for the nearby Leine River to discuss once more, this time on site, the options for diverting water for the Herrenhausen water features. Again Leibniz lays out his idea for a canal. Also present are a fountain expert and an architect. The latter points out that there needs to be enough of an incline to give the water "lots of rush."[52] Helmont has been asked to come along because he has some experience with technical and mechanical matters and Leibniz values his practical expertise. The metaphysics of animated nature isn't the only link between the two men; both have a mind for the practical and a desire to promote the general welfare through concrete endeavors. In this shared ambition to improve the world holistically, Leibniz sees a deep spiritual kinship between Helmont and himself.

After the conference by the river, the group splits up, and Leibniz heads back into town. In the evening he thinks back on the day and sets down what happened in his newly begun diary. Once again he has experienced, thought, and written so unbelievably much in the span of twenty-four hours that today we have a hard time ascribing all he's accomplished to just one person. Even his contemporaries found it hard to fathom such a dense concentration of activity. Bernard Le Bovier de Fontenelle, for example, when delivering his eulogy of

Leibniz in 1717, a year after his death, will decide to divide up his life in the telling, as though he were narrating the lives of several scholars.[53] Today it would take several people to collectively master the various scientific disciplines in which Leibniz was equally at home.

How then can we be sure that the events and actions described in this book all really pertain to one and the same person? Ultimately we have no choice but to trust the author (that is, me) when he asserts that all of the sources brought to bear actually concern this particular historical figure. And what about Leibniz himself? Is the person who on the morning of August 13, 1696, receives word of his promotion to Geheimer Justizrat identical with the one who at midday, at Herrenhausen, reads a letter from Versailles, and also with the one who goes out to the Leine with Helmont in the afternoon? For Leibniz, the question isn't hard to answer; in his mind the events are conjoined to form one unity of experience. Memory takes on an important function, supported in turn by the diary.

According to Leibniz's own thinking, the actions of memory—that is, the interplay between remembering and forgetting as part of reflexive consciousness (apperception)—are critical in enabling a personal identity to manifest and maintain itself.[54] But Leibniz—in contrast to, say, John Locke—is not of the view that what constitutes a person is solely dependent on a consciousness that joins the various mental states together. Gaps in memory that occur due to illness could be filled by what others relate. Memory loss, moreover, doesn't exempt a person from punishment for crimes that they might have committed before the loss occurred. Thus, according to Leibniz, recourse to consciousness alone is not enough; more is needed if one is to characterize oneself or another as a person and to ascribe to that person actions for which they can be held responsible.

For Leibniz, *something* must explain why our consciousness links together a series of states to form a single unity; there must be a sufficient reason. Further, an abiding active principle is needed that produces these states and coordinates their sequence. They cannot be regarded as independent and isolated. Leibniz explains this necessity by using as an example the theological debate over transubstantiation. In his view, during the eucharist a piece of bread cannot be replaced by the body of Christ, not in the sense that holds that the qualities of the bread—what a philosopher would call its accidents—being independent entities, pass from one object to another. Qualities like brownness and squareness exist not in and of themselves but only as properties or conditions of something. By this same logic, states of consciousness are not autonomous either; rather, they point to the existence of something that precedes them, something upon which they are predicated. Neither implanting the single unity formed by one person's consciousness into another carrier nor splitting up that unity into two brains is possible without the person becoming one or several others.

What this means for Leibniz is that there has to be a substance that ensures that a human individual—that is, a being with mental faculties, a person—constitutes a single unity (of plurality): the same simple, indivisible substance that Leibniz, beginning in the summer of 1695, hesitantly calls a monad. This unity is utterly distinctive, unique in all its particulars, meaning that even the slightest alteration to a person's past actions would result in that person becoming another, different person. An Adam who lived in another possible world in which he hadn't sinned and hadn't been driven out of paradise along with Eve would be a completely different Adam.[55] Leibniz (unlike, again, a thinker like Locke), does not content himself with a purely psychological answer to the question of what a person is. But there remains the problem of ultimate justification— what is the cause of the cause?—and this he can solve only by biting into the "theistic apple."[56]

LIFE-FORMS IN FLUX

What emerges from the brief hints in the diary on this August 13 is a world full of thoughts on fundamental questions that today are primarily the remit of the so-called life sciences, ranging from biology to psychology. But we should not to be too quick to link Leibniz to these disciplines. "Life," for him, is not defined, as it has been since the nineteenth century, using criteria like organism, organic functions, reproduction, anatomical structures, and metabolic processes.[57] For him, perception alone, attached to an organized substance (*substance organisée*), is a sufficient criterion for defining a thing as alive. With this view, Leibniz will later oppose the vitalistic theory of Georg Ernst Stahl, according to which there exists an independent life force that marks the difference between organic and inorganic matter.[58] It's possible, still, that this prebiological concept of life may be of interest to us as we continue to question and problematize the distinction between organic and material substance, between living and nonliving matter.

Leibniz will take up the subject of life and living things one more time on this Monday in August 1696. He writes a letter to the historian William Ernst Tentzel that goes unmentioned in the diary. Toward the end of the previous year, Tentzel studied a giant skeleton that had been found in a sand pit near Gotha, and by making anatomical comparisons with other animals, he had come to the conclusion that the fossilized bones had belonged to an immense elephant. For a few weeks now, the fossil discovery has been the subject of a lively exchange of letters between Leibniz and the scholar from Gotha. They are in agreement that the bones must be the remains of a real living creature and not, for instance, purely mineral, inorganic objects, as the expert commission assembled by the Duke of Saxe-Gotha-Altenburg claims. Leibniz considers Tentzel's conclusion to be entirely plausible, but in his letters he repeatedly argues that the bones could also be the remains of a sea creature related to the walrus.[59]

The issue at hand isn't what life is but rather how life might have changed over the course of history. With his marine animal hypothesis, Leibniz is pointing to his theory that the North German Plain was underwater at some point in an earlier epoch. At the same time, he is reflecting on whether the appearance and nature of living things might have changed over time. In doing so, he comes dangerously close to the limits of what can be said in his era, when the church-sanctioned dogma still holds that animal and plant species have persisted, unvarying, since the time of creation. To claim that species undergo changes or even go extinct is strictly taboo. To posit this notion would be blasphemy; it would contradict the divine plan for creation. And so Leibniz is careful only to touch on the delicate subject, for which others have been burned at the stake. "I don't claim that certain *species have gone extinct*, though I won't go so far as to say this is absurd," as Leibniz writes to Tentzel on August 13, "but I do think that we must make a distinction between species that have gone extinct and those that have drastically changed. Thus dog and wolf, cat and tiger can be seen as belonging to the same species. The same can be said about the amphibious animals or the sea cows in relation to the elephant."[60] To assert that change occurs within species still conforms with the prevailing doctrine, but nevertheless, by writing and suggesting what he does here, Leibniz is going way out on a limb. Elsewhere in his writings on natural history, he will be even more explicit and go considerably further. And when he does, he will give shape to ideas that anticipate by a fair margin the nineteenth-century theory of evolution as developed by Charles Darwin and others (see chapter 7).

MONADS 2.0

And so a long day comes to an end. Leibniz sets down his quill, satisfied. All right, he's still got it. The day was packed to the brim with activities, back to back to back, with not a minute wasted, just the way he likes it. And on top of that, he got the most important stuff down in writing. It's been a good start for his push to drive off exhaustion and apathy. No doubt he's going to continue keeping the diary he began today. The entries he will make over the next few weeks will tell a lot about how Leibniz spends his days: his conversations with Helmont, the task of planning the canal for the garden at Herrenhausen, and the visitors he receives. He describes how he sits up late into the night poring over documents and chronicles for the history of the Welfs, notes the letters coming in and going out and his frequent trips to Celle and Wolfenbüttel—and finally, more than once, he recounts lingering in bed in the morning while the solutions to difficult problems involving curves and tangents come to him. Bit by bit, Leibniz seems to recover his strength and gradually get back to feeling like his old self again. Malaise and gloom are shaken off, or better yet: worked off. To assess how much he has accomplished, he needs only to flip back through his daily notes. And so, little by little, the crisis is overcome.

The necessity of keeping a diary thereby grows less and less urgent. In September the entries get shorter and less substantial. The more his self-assurance grows, the less significance his written self-observation has. On October 10 the by-now-scanty notes come to an end. In early 1697 he writes a few scattered entries, then sets aside the diary for good. Leibniz has successfully concluded his self-prescribed writing cure, and in the meantime so many new tasks and projects have come up that he has no time left for it anyway. The workaholic is at peace with himself once more—content with the sedative of steady activity. The reactivation of the term *monad*, which up to now he has kept under wraps, is another sign that Leibniz has recovered his confidence and self-assurance. Almost fourteen months after its first appearance (in the draft of a letter to L'Hôpital), the term reappears in a letter to the Italian mathematician Michel Angelo Fardella dated September 13, 1696[61]—that is, at the very point when the diary is becoming less necessary. True, Leibniz is still a bit unsure as to whether the term precisely captures what he means to convey in his metaphysics of substance. But the term won't disappear again; to the contrary, he will use it more and more, including in print, and soon enough it will become the central concept of his metaphysical program. Leibniz has rediscovered the pleasure he once found in philosophy and science. August 13 serves as an example of how he is capable of using self-healing methods (writing as self-reflection, piling on the activity as a means of overcoming a lack of motivation), to emerge from crises stronger than before. What some today call resilience is for Leibniz a resource that, thanks to his philosophy of optimism, can always be renewed. Indeed, for him it seems inexhaustible.

CHAPTER 4

BERLIN, APRIL 17, 1703

DIVIDING THE WORLD INTO ONES AND ZEROES: PATHS INTO THE DIGITAL FUTURE

Day in, day out, day in, day out, day in, day out . . .

JOY DIVISION, "DIGITAL," 1978

IN SEARCH OF A GLOBAL FORMULA

Today is a special day. A day in which history will be made—though not with any great fanfare; what happens will happen quietly and go unnoticed. The date (according to the new calendar) is April 17, 1703. Leibniz believes he has found important clues that point to a way to transform nonnumerical information into mathematical symbols. It's not a breakthrough, not a revolutionary beginning—but still, Leibniz thinks he has come one small step closer to his great aim of making the world and its knowledge accessible via calculation. It's one step farther along the road toward a universal method that would make it possible to break concepts like "wind," "fly," or "trinity" down into simple signs and reckon with them as one would numbers, thereby gaining new knowledge. On this day, a Tuesday in a fitful spring, Leibniz pens a letter to the Irish scientist

Hans Sloane. He reports to him that a Jesuit priest living in China has discovered how to use Leibniz's binary number system to decipher an ancient Chinese system of writing and symbols: "Thus we have the solution to a mystery that the Chinese have lacked knowledge of for more than one thousand years and that, having lost its true meaning, has begotten ignorant interpretations."[1]

Key aspects of our modern digital culture can be traced back to Leibniz's theoretical efforts in combinatory analysis and in pursuit of a universal language of symbols (see chapter 1). Today life without computers, the internet, email, or social networks is scarcely imaginable. Without a doubt, the future belongs to the digital. Intelligent machines, the internet of things, and quantum computing are probably mere forerunners of a still largely unknown era. But when did it all begin? Where lie the origins of the universal culture of zeroes and ones?

The history of digital data processing has not followed a linear course; nor can we say exactly when it began. Only in hindsight can we see how meandering the path toward digital modernity was and recognize distinct stages and landmarks along the way. Leibniz is certainly part of the story. True, many of his ideas were vague and speculative, came to nothing, or were later developed by others independently of him. Looking back, however, a single image comes together, formed of what for him were loose, unfinished elements, even if sustained by a comprehensive vision: that the world is calculable, that knowledge can—literally—be reckoned with. Only in hindsight is it possible to recognize that these elements are like little tiles forming part of an ever-shifting mosaic. Leibniz contributed at least *four* of these "tiles." The fourth reveals itself on April 17, 1703.

SUMMER OF LOVE

Leibniz writes to Sloane in London not from Hanover but from Berlin. For more than five years now, he's been shuttling between

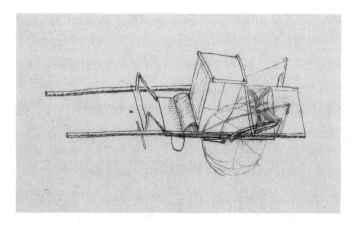

Gottfried Wilhelm Leibniz, drawing of a mountable cabin unit for an open post-chaise, c. 1700.

the two cities. From his base on the Spree, he undertakes longer excursions to Dresden, Prague, and Karlsbad (today Karlovy Vary), once even making a secret trip all the way to the court of Emperor Leopold I in Vienna.[2] The rolling stone continues to seek out opportunities to travel; he'd rather be constantly on the move than forever in one place. From his diary of 1696, we know that good ideas often come to him while he is traveling. If only there weren't all those potholes, those uncomfortable seats, and the constant jostling of the coach. Leibniz has a one-person cabin built that can be mounted on an open post-chaise. The cabin is covered in a waxed linen sheet, can be moved from chaise to chaise, and has a seat outfitted with a suspension, so that he can read and write with minimal disturbance as he rides. Travel fuels his inventive spirit, and many of his inventions are geared toward improving the experience of travel itself. The idea to fit out a carriage with dual wheels doesn't make it past the draft stage, however, and his mobile cabin for post-chaises doesn't catch on. Still, it would be a major advance to be able to travel faster. Johann Joachim Becher, who is involved in his own share of eccentric projects, pokes fun of Leibniz at one point, saying Leibniz wanted to have a coach built that could make the trip from Hanover to Amsterdam in six hours.[3]

Leibniz refers to his mobile litter with affection as his "post-chair," and bouncing between Berlin and Hanover, he finds himself between seats. In Berlin, seat of the Prussian king Frederick I, he has acquired the title of Geheimer Justizrat, the same title he holds in Hanover. By the time he makes his last trip in 1711, he will have traveled to Berlin more than a dozen times, usually staying for several months each visit; all told, he will spend more than three years of his life there. He doesn't want to give up his position in Hanover, however. Occasionally he considers making the switch and entering fully into Prussian service, but he can't bring himself to take the final step. In the up-and-coming Prussian capital, Leibniz finds plenty of fellow intellectuals to converse with, puts his efforts into advancing the reconciliation talks between the Protestant and Reformed churches, and sets about founding a scientific academy. Many dreams seem to come true at the royal court in Berlin. Above all, Leibniz feels drawn to the educated and attractive Queen Sophie Charlotte, the thirty-five-year-old daughter of Electress Sophie. She calls herself his pupil and keeps inviting him back to Berlin.[4]

As of April 17, 1703, he has been living in this city of about thirty thousand for more than ten months. He spent the balmy summer of the previous year outside Berlin at Lietzenburg Palace (also known as Lützenburg Palace). He looks back fondly on this time: there was a great deal of laughter and reveling at the pleasure palace of the Prussian queen. A motley company of noble courtiers, clergymen, scholars, and artists gathered for informal entertainments, lively discussions, and lengthy strolls in the park. The guests discussed the latest books in the daylight hours, then danced the night away. One of them was John Toland, who in England had been denounced as an atheist. With him, Sophie Charlotte and Leibniz discussed the provocative theories of Pierre Bayle—whose *Dictionnaire historique et critique* everyone was talking about—concerning the division between faith and reason. The queen enjoyed watching the young Irish free spirit and the somewhat stiff German scholar butt heads and exchange intellectual blows, egged on by the illustrious company. The entire court enjoyed this riotous mixture of sublime metaphysics and mundane male posturing.[5]

Leibniz and Sophie Charlotte—without a doubt, the intimacy between them was strong. There has been much speculation about the nature of their relationship. Was their affection for one another really just platonic? The divide between the aristocratic—indeed royal—lady and the burgher-class scholar seemed insurmountable; it stood between the two of them like an invisible wall, even in the carefree hours at the summer palace and in its gardens. It's fair to ask of course whether the palace guests were serious about defying societal constraints. Was it radical enlightenment or harmless flirtation with the outré? In any case, the Lietzenburg celebrity culture had no trouble celebrating itself. Toland claimed that cannibals were purely an invention of the Spanish, who were looking to justify the cruelty of their politics of conquest. Electress Sophie, who was spending this summer as a guest of her daughter, scoffed that if he continued trying to discredit the Christian religion, the cannibals would soon be the only ones listening to him.[6] The nobility, a class that is growing ever more wealthy, is well on its way to unmooring itself from the rest of society, and Leibniz, with no noble title of his own to boast of, is made constantly aware of the many fine distinctions between himself and his social superiors.

Be that as it may, better to meditate on the ephemerality of the world with cake and champagne than without. By the time he experienced this fairy tale Berlin summer of 1702, Leibniz had long since overcome his existential crisis of 1695–96. Highly respected in the European republic of letters, he was a regular guest of the powerful. He too enjoyed the moments of idleness and excess that were available to him. But they quickly got to be too much; lurking behind satiety and contentedness, he sensed the seeds of decadence and decline. To reach every goal is to have nothing left to strive toward. Just a short time after his arrival at Lietzenburg, he wrote: "It seems that all too great comfort is not good, in that it causes people to waste their life along with their time without noticing, so to speak, and neither use it sufficiently nor feel it."⁷ *Ex bono malum*: an excess of good can also have bad results. Being close to power is seductive, but before one gets sluggish, it's better to grab the next post-chaise and rattle on out of there. Leibniz was able to bear becoming part of the establishment only by being forever on the go. In his constant travels between courts, the coach remained his true home.

ON BRÜDERSTRASSE

After fall came and the merry summer crowd left Lietzenburg, a harsh winter ensued, and Leibniz suffered a leg injury that now threatens to become chronic. "All winter long I have struggled with complaints that the autumn dealt me," he writes to Sloane on this Tuesday in April 1703.⁸ From Hanover, the fifty-six-year-old receives a steady stream of grievances from his servant Johann Barthold Knoche (referred to in the letters as his "Vallett"), who complains of being short on money and having trouble procuring wood. A few days before April 17, Leibniz learns that Knoche soon won't have anything left to wear; his tattered clothes are literally falling off him.⁹ No good news from Hanover, then; this shuttling between the two cities is starting to feel like being torn in two.

Leibniz hasn't seen the queen recently. He writes her that even while his health is holding him back, he has been having philosophical discussions on the relationship between freedom and fate with his neighbor in the building, François d'Ausson de Villarnoux, who works for Sophie Charlotte as chief equerry and chamberlain.¹⁰ Since Easter (April 8–9) he has been doing a bit better and has been able to leave his quarters again from time to time. Leibniz is probably living in a small rooming house close to the palace on "Brüder Strasse," the preferred spot for court officials, among them Friedrich Ernst von Kniphausen, the chamberlain to the Prussian king.

The street, which owes its name to a medieval Dominican mon-

Portrait of Leibniz after
Andreas Scheits, oil on
canvas (copy), 1703.

astery, is among the most sought-after places to live in the twin
cities of Berlin-Cölln. Over the course of the eighteenth century, it
will be home to significant figures in the city's cultural life like the
writer Friedrich Nicolai and the demographer Johann Peter Süßmilch. On Brüderstraße, Leibniz can linger in bed undisturbed
if he's so inclined, dedicating himself to his thoughts and reading
and writing extensively. A new "night gown" given him by Andrew
Fountaine provides the necessary comfort.[11] Meanwhile Electress
Sophie imagines Leibniz in Berlin, still lying in bed and drinking coffee, while she, in Hanover, likewise still in her bedchamber,
enjoys her breakfast cocoa. Naturally she can't help but get a dig in
about the dangers of overindulging in coffee.[12] But in the days leading up to April 17, Leibniz is ever more on the mend.

LIVE TRANSMISSION

Leibniz has got a fair amount on his plate this Tuesday. Six letters
in his hand, some of them quite long, can be traced to this date,

whether directly or indirectly. In addition to Sloane, the addressees
of these letters are: Pierre de Falaiseau, who lives in London; Johann
Georg Eckhart, Leibniz's assistant and secretary; the Jesuit priests
Joachim Bouvet and Jean de Fontaney; and finally Christophe Brosseau, the Hanoverian Resident (ambassador) representing the Electorate of Brunswick-Lüneburg in Paris.[13] If we include the letters
that Johann Fabricius (the younger), Alphonse des Vignoles, Carlo
Maurizio Vota, and Eckhart all write to Leibniz on this date,[14] as
well as the odd letter or two whose date can be established only as
being sometime around mid-April, then this Tuesday undoubtedly
numbers among the days with the highest volume of letters written
and received by Leibniz in 1703. Leibniz's own letters travel in every
direction, to England, Paris, and Poland, and from thence over the
sea, around the whole of Africa, and on as far as China. Several letters are bundled together to be sent all at once. Some survive only
as drafts or excerpts copied out by Leibniz. The existence of others, in turn, can be proved only indirectly through references in the
surviving correspondence. The drafting of a letter can extend over
protracted periods of time. The lengthy letter to Bouvet, for example, Leibniz has worked on in various stages, making corrections
and changes at every new juncture, finally having a fair copy made
from the undated draft and adding a postscript. The extant sources
indicate that the seventeenth was the day on which the letter was
completed.[15] And even if Leibniz did not in fact send the Bouvet
letter off this Tuesday, its contents are still closely connected to the
letter to Sloane bearing this day's date in Leibniz's own hand.

    Leibniz's correspondence crosses borders, collapses distances,
and stretches a single conversation out over months and years. How
protracted such a conversation can get is demonstrated by the letter
from Joachim Bouvet, the Jesuit priest, which Leibniz answers over
the course of several days this April: it has taken almost a year and
five months to make it from Beijing to Berlin. Leibniz uses various
channels and postal hubs for his correspondence. Guiseppi Guidi,
who as secretary to the elector in Hanover maintains a veritable news

bureau, acts as one of these collection points. He sends letters clear across Europe, disseminating the latest news, and supplies Leibniz with a fresh batch once or twice a week. When communicating in such a manner, certain precautionary measures are indispensable. Leibniz, who maintains ties to several princes' courts, must be careful what he writes to whom. Encrypted words, false addresses, postal detours, inconspicuous seals, and pseudonyms are all part of his regular repertoire for communicating by letter or parcel.[16]

Sometimes he refers to certain persons only vaguely, as *ami*, or uses codenames. In order to conceal the fact that he intends to introduce the cultivation of silkworms, which he'd originally planned only for Berlin, simultaneously in Saxony as well, he assumes the fictional persona of "Monsieur Kortholt" to write his letters, using the name of an actual person who at the time is imprisoned in debtor's jail in Berlin for embezzlement.[17] Leibniz's mania for disguise goes so far that on seeing a newly finished portrait of himself, he responds with the comment, "A copperplate engraving prevents a man from remaining incognito." (Sophie meanwhile finds that the portrait depicts him with a fat drinker's nose.)[18] Still, he's by no means paranoid. Both in Hanover and in Berlin, he is repeatedly accused of spying for the other side—though both are part of the Holy Roman Empire, the Electorate of Hanover and the Kingdom of Prussia remain wary of each other. When diplomatic relations between the two states turn sour at the start of the year, he discovers on several occasions that his mail has been opened. The suspicion of espionage also eats at the queen. This spring Sophie Charlotte sends her court doctors to Brüderstraße not just out of legitimate concern but also to check whether Leibniz is actually sick.

REPORT FROM EUROPE

The spectrum of topics in the letters written on April 17 is broad. Almost every letter consists of a wide variety of news: personal circumstances, gossip from the different courts, the latest from the scholarly world (*nova literaria*), and, if it's an official he's writing to, organizational matters. Alongside the exchange of information, another practice common to *commercium literarium*, as correspondence among scholars is called, is to forward enclosed letters or short messages (known as *billetts*) to other recipients. Journals and books that are difficult for the recipient to find locally are also included. The actual contents of the letters are not always of prime importance; the correspondence often has a strategic function as well. Leibniz is addressing Sloane, for example, in his role as secretary of the Royal Society in London. Now that Leibniz is president of the Berlin Academy of Sciences, which he founded in 1700, he can correspond with his English colleague on equal terms.

Leibniz still has his hands full trying to get the young society off the ground. It's short on members and funds. Money is supposed to come primarily from the sale of calendars, on which the academy is to receive a monopoly. Leibniz is also thinking of other potential sources of revenue, like the silk production project, lotteries, or the implementation of a new kind of fire hose (*Brandspritzen*). The latter is meant to demonstrate that his aim isn't ivory tower scholarship but rather application-oriented science; the goal is "to unite theory and practice."[19] More than anything, Leibniz intends to fiercely promote his academy. He tells Sloane about the still unknown astronomical observations of comets that Gottfried Kirch, a founding member, made back in 1686–87—and he proudly and confidently calls Kirch "our observer in Berlin" (*observator noster Berolinensis*).[20]

It's astounding how Leibniz can concentrate on so many details pertaining to such totally different fields when at the same time he's got so much information coming at him. Already piled up on the desk are Guidi's letter from Hanover; a message from the queen that has probably just come in by speedy royal courier; and from Marienburg in Poland (present-day Malbork), a letter from

Reichsgraf Jakob Heinrich von Flemming, who writes with a question. All of it is important and urgent, naturally. What should he reply to first, and what can wait? Which subjects overlap between the letters and recipients, what can he combine—but more important, what must he absolutely not write to whom? No sooner has the enmity between the Prussian king and the Hanoverian elector (over the occupation of the city of Hildesheim, which is allied with Berlin, by the Duke of Brunswick-Lüneburg's troops) subsided than the two courts again lock horns, this time on account of an Italian monk named Attilio Ariosti, whom Sophie Charlotte is absolutely intent on keeping in Berlin to serve as a musician in the court orchestra. And Father Carlo Maurizio Vota is still deeply upset about the breakdown of discussions between Reformed and Protestant theologians in Lietzenburg.

Once, possibly this very month, Leibniz comes close to making a potentially grave mistake: caught up in the flow of his writing, he forgets the codename that he and a correspondent had agreed upon. While writing a letter, the words "I wish" slip from his quill. Leibniz spots the *I* that could give him away and corrects it: "M. Cortholt wishes."[21] There's no thinking what would have happened if Sophie Charlotte, who reads the letters he sends from Brüderstraße (out of politeness to her, he leaves them unsealed), had found out in this way that he is seeking to realize his plans for silkworm cultivation—plans that are meant to be exclusive to Brandenburg—for the rival Electorate of Saxony as well.

Everything is overshadowed by the calamities of high politics. Dark clouds lie over Europe. For three years now, the Nordic War for dominance in the Baltic has raged between the Russian czar Peter I and Sweden under King Charles XII. And since the autumn of the previous year, the War of the Spanish Succession—after initial skirmishing between the chief adversaries: France on the one side; England, the Netherlands, and Leopold I and the Holy Roman Empire on the other—has entered its bloody stage now that the Bavarian elector has pulled out of the alliance against Louis XIV and invaded the Palatinate. Even the notoriously optimistic Leibniz sees little grounds for hope at the moment. In his long letter to Bouvet, he concludes his gloomy report on the state of war-wracked Europe with the words "As for European affairs, they are in such a state as to make us envy the Chinese."[22]

LEARNING FROM CHINA

If today Leibniz turns his gaze away from crisis-ridden Europe and toward China, it's not just trade relations that are on his mind (such as the silver trade); he's interested in a comprehensive exchange of knowledge with the Middle Kingdom. For many years, he has cultivated contacts with various Jesuit missionaries in China. In addition to Bouvet and Fontaney, there is also Claudio Filippo Grimaldi, whom Leibniz met in Rome in the summer of 1689, shortly before his departure for Beijing. In 1697 Leibniz published a selection of their letters and writings under the title *News from China* (*Novissima Sinica*), and ever since then, he has been regarded in Europe as a certified China expert. To Sophie Charlotte, he once wrote that he was going to have a sign mounted on his door with the inscription "China News Bureau," so that everyone would know to whom to turn for the latest information on the subject. Sloane also knows about Leibniz's sinophilia, which is why in his last letter he enclosed a copy of an account written by an English doctor in China with much information about life in the country.[23] Leibniz regards Chinese culture as equal to European culture, and in fact he sees the distant kingdom, on the far side of the Eurasian continent, as the cultural counterpart to Europe. From the "Anti-Europe," he expects above all knowledge that is technologically and practically oriented. In return, he believes the Far East will be interested primarily in the West's theoretical knowledge, especially in mathematics. Again and again, however—including in his most recent letters to Bouvet and Fontaney—he warns the Jesuits not to reveal too many teachings to the Chinese, or else European civilization might fall behind.

These days the China News Bureau has its headquarters on Brüderstraße. He asks Eckhart to inquire, should he travel to Moscow, about the trade caravans traveling between Moscow and Beijing.[24] And he reminds Bouvet and Fontaney that he has been waiting a long time for answers to the list of questions he sent them. He's interested in Manchurian and Chinese dictionaries, geographic information, hints for determining the lines of longitude that are so important for seafaring, medical remedies for smallpox or syphilis, and then of course the secrets of Chinese silk production—plus the manufacture of fine paper and some of the special inventions that he's heard of, such as airtight containers of leather-like material that can be used for sleeping on.[25]

DIGITAL DREAMS

But at the moment something else is driving Leibniz's interest in China. In his last letter, Bouvet produced a link between the binary numeral system that he had learned of from Leibniz and a millennia-old collection of Chinese oracular sayings. Leibniz refers to this at length in the letter he is writing now. He believes Bouvet has discovered something that even in his wildest dreams he couldn't have imagined. Before he elaborates any further, he rehashes once again how more than twenty years earlier he developed his system of arithmetic using 0 and 1.[26] And sure enough, Leibniz's preoccupation with this peculiar method of calculation can be traced back to his time in Paris. In contrast to the decimal system, the base-2 numeral system—Leibniz speaks of a binary or dyadic system—has only two numerals. Leibniz converts decimal numbers into binary numbers by expressing them each as the sum of exponential products of two:

Decimal system	Sum of exponential products of two	Binary system (Dyadic)
1	1×2^0	1
2	$1 \times 2^1 + 0 \times 2^0$	10
3	$1 \times 2^1 + 1 \times 2^0$	11
4	$1 \times 2^2 + 0 \times 2^1 + 0 \times 2^0$	100
5	$1 \times 2^2 + 0 \times 2^1 + 1 \times 2^0$	101
Etc.	Etc.	Etc.

Accordingly, the first ten numbers of the decimal system are given in the dyadic numeral system as follows:

Decimal System	Dyadic System
0	0
1	1
2	10
3	11
4	100
5	101
6	110
7	111
8	1000
9	1001
10	1010

Here is another demonstration of Leibniz's characteristic facility with symbols and their reinterpretation. The sequence of digits "10" does not symbolize ten; rather, it is read as "one-zero" and is assigned the value of two in the binary system. Using this system, the four basic operations of arithmetic can also be carried out. Here, for example, is addition:

1	1
10	2
100	4
111	7

Leibniz is neither the first nor the only mathematician to experiment with binary numbers. But in his letters and writings, he continually promoted this numeral system, so much so that today his name is inseparably linked with the development of binary arithmetic, out of which grew the digital code used in computer technology. In 1938, before building the Z3, one of the first functional computers, Konrad Zuse, a pioneer in the field, will use the term *dyadic* for the mathematics behind his trailblazing circuitry, and in so doing he will make explicit reference to the work of Leibniz.[27] The binary calculation method is only *one* milestone on the path to the modern digital age that can be linked to Leibniz; a *second* is the use of this method to program a machine. The binary or dyadic numeral system not only has the advantage of logical clarity (the digits *1* and *0* correspond to "true" and "false") but also offers the benefit of being easier to work with in a technical regard.

In 1679–80 Leibniz came up with ideas for two calculators that would use the binary system. One used marbles that were supposed to roll along grooves and drop into holes, while the second had number wheels and number converters (for switching between decimal and binary numbers). Unlike the decimal calculator developed in Paris and its successor models, the two dyadic or base-2 machines remained designs on paper. Similar to the machines that worked with decimal numbers, however, they were supposed to be equipped with mechanisms for saving intermediate results— the marble model, for example, by storing those marbles that had fallen through one of the holes. When carrying out a binary addition operation—say $1 + 1 = 10$—one marble would slide into a hole that corresponded to 0, and the second marble, the 1, would keep rolling. The same principle lies behind modern digital calculators, namely a dynamic alternation between two stable states: only here the movement between "on" and "off" is triggered not by electrical impulses but by the mechanical interplay between "marble there" and "marble not there."[28]

Leibniz has been thinking about how the binary principle could be used in algebra as well. Multiplication with binary numbers, for example, yields four possible outcomes: $0 \times 0 = 0$, $0 \times 1 = 0$, $1 \times 0 = 0$, and $1 \times 1 = 1$. Put in logical terms, if "true" $= 1$ and "false" $= 0$, then these outcomes correspond to the four possibilities 0 and 0 = 0; 0 and 1 = 0; 1 and 0 = 0; and 1 and 1 = 1. If one replaces the numbers with letters, as is common in algebra, then the last of these possibilities can be represented as: a and $a = a$. In order to express this maxim, Leibniz creates a symbol, a plus sign inside a circle: $a \oplus a = a$.[29] Later George Boole, independently of Leibniz, will present this maxim as $xx = x$ and on this basis develop an algebraic logic that uses the operators "and" (conjunction), "or" (disjunction), and "not" (negation), familiar to us today from computer programming. With their help, the logical values 0 and 1 can be linked together to form expressions whose complexity can be increased almost at will.[30] Given all this, we could say that Leibniz managed to contribute to *three* crucial components of modern digital technology: binary numbers (mathematical "mosaic tile"), dyadic machines (hardware "tile"), and binary algebra (logical "tile").

From Leibniz's point of view, these components all line up with those he needs to realize his chief ambitions for a general theory of knowledge based on combinatorial analysis, a universal symbolic language (*characteristica universalis*), and a general calculation method for arriving at conclusions (*calculus racionator*): simple symbols (0 and 1), logical rules to guide operations with them, and ideas for automated processes (like the calculating machine with its marbles, grooves, and holes). Leibniz has already taken initial steps in each of these directions. So far, so good. But what's still largely missing are concepts and methods for converting nonnumerical information into numerical values. Words have meanings, refer to things or facts, express a sense. Numbers are meaning-neutral. So how can concepts be converted into numerical signs that can be plugged into calculations that yield something that makes sense, something with meaning attached to it? Or to put it in the language of modern information theory: in what way can semantic figures be transformed into syntactic structures—and vice versa—and subjected to automated calculation methods? This has remained an unsolved problem for Leibniz, at least until spring 1703. Now, though, there seem to be signs of movement.

A WORLD OF ZEROES AND ONES

In his recap for Bouvet, Leibniz also goes into the mathematical uses, in a narrower sense, that he hopes to get out of the dyadic. He doesn't see many advantages for everyday use (with the exception of weighing coins), but he does perceive benefits for number theory. What's important are the regularities that can be found in series of numbers like square numbers, cubic numbers, and other exponential values of natural numbers. In dyadic number series, periodically reoccurring patterns of ones and zeroes are clearly recognizable—more so than the patterns in series of decimal numbers—and this makes it easier to analyze the arithmetical progressions of these series. Moreover, Leibniz expects to glean important insights from the binary system into dealing with difficult values like pi or with arithmetic involving fractions.[31] He complains that he lacks collaborators to help with the necessary calculation work. Still, the founding of the Berlin Academy of Sciences now offers hope. In its orbit he can seek out qualified mathematicians, like Philippe Naudé and Pierre Dangicourt, whom he has put to work carrying out dyadic investigations and test calculations.

In Leibniz's binary thought, mathematics has been tightly bound up with metaphysics from the start. In this sense, he maintains a strong connection to the tradition of Platonic and Pythagorean philosophy.[32] God created the world by measure, number, and weight, and Leibniz, with the help of his dyadic calculation method, believes he will henceforth be able to define this belief more precisely: 1 represents the unity or the one, 0 stands for nothingness or the lack of existence, the void. Consequently, God, as the absolute unity (1), created the world out of the void (0). The void is dispelled by creation; conversely, creation is bounded by the void. This means that nothing in the world is consummate, nothing is whole or perfect, but at the same time there is no *absolute* nothingness, only *relative* nothingness. For this reason, Leibniz firmly rejects the idea of an absolute physical vacuum postulated by Newton.

All creatures differ from one another in their individual mix of divine perfection (1) and nothingness (0), which can be represented by short or long chains of ones or zeroes, chains that also depict the hierarchy of creatures. The series that underlies a human being, one could imagine Leibniz saying, is longer than that of a fly but shorter than that of an angel. In an analogy drawn from geometry, Leibniz compares this hierarchy to circles of various sizes, each of which has its proper limit defined by its circumference.[33] However long and complex these figures made of zeroes and ones might be, with their basic two-value structure they are yet another manifestation of the elegance and beauty of the universe, which results from the unity of plurality. The dyadic is for Leibniz nothing less than the symbolic expression of a world of diverse abundance and, at the same time, sublime clarity. Just as every single leaf in a garden in turn contains another garden within itself—just as the world is overabundant with living things—so are numbers overflowing with other numbers. There are no gaps or leaps, only steady transitions. The small becomes ever smaller, but it never becomes nothing. What applies in infinitesimal calculus holds true for all of nature: the wave breaking on the shore wouldn't produce any noise if it were composed of nothing but little nothings.

Nothingness (*nihil*), however, is everywhere. For Leibniz, thinking in terms of a logical formula, this means A + nihil = A. Because nothingness can be added to any given concept (without changing it), that means that it is present in any given concept.[34] Nothingness, then, is nothing to be afraid of. Leibniz counters the *horror vacui* of antiquity—the fear of the void—with the productive power of zero. Zero then is not nothing; rather, mathematically speaking, it is the twin of infinity. No longer does it merely stand for the absence of a number; rather, it is recognized as a number in its own right.[35] In order to be able to count apples in a bowl, at least *one* apple has to be in the bowl. If, however, there are *no* apples in the bowl, it doesn't mean I've miscounted, only that I need to put more apples in the bowl. Leibniz attributes other positive functions to zero as well: for example, because zero represents lack, it ensures that we strive toward the good. And the idea that there is no absolute nothingness in the world correlates with the Leibnizian notion that there cannot be any real death in it, either. Thus Leibniz's dyadic, seen as a kind of worldview, is his metaphysical optimism in mathematical garb.

A CHINESE ORACLE IN BINARY STRUCTURE

Leibniz is so excited about the correspondence between binary mathematics and the Christian belief in creation out of nothingness (*creatio ex nihilo*) that in 1697 he proposes to Duke Rudolph Augustus in Wolfenbüttel that he have a commemorative coin minted depicting the creation of the world (represented by the night sky, sun, and stars) and a pyramid made of ones and zeroes.[36] Leibniz also sees the correspondence as being particularly well suited to missionary work, including that of the Jesuits in China. The notion that

God created everything out of nothingness without making use of any kind of original matter was, he felt, hard for non-Christians to understand. If one were to show them how all numbers can be formed out of combinations of 1 (unity) and 0 (nothing), Leibniz writes to Bouvet on February 15, 1701, then it would also be easier to convey the mystery of creation, especially to the Chinese, given their interest in mathematics.[37] Bouvet approves of this idea. In Beijing, the Jesuit missionary is just then occupied with his effort to decode the *I Ching* (*Book of Changes*), a collection of oracular writing dating back to the eighth century B.C.

The *I Ching* was consulted with the help of broken and unbroken yarrow stalks, later represented by divided and undivided lines. The oracular text was composed of either three or six lines, usually in the form of eight groups of three (trigrams) or sixty-four groups of six (hexagrams). The eight trigrams were held to be fundamental characters, each corresponding to a natural element: sky, earth, thunder, water, mountain, wind (wood), fire, and lake. The text was believed to date back to the legendary Emperor Fu Xi. In seventeenth-century China, however, no one knew anymore what the individual hexagrams and their corresponding characters meant exactly.

But when Bouvet is introduced to Leibniz's binary mathematics, what he believes to be a correspondence between the unbroken yang and interrupted yin lines on the one hand, and the digits 0 and 1 on the other, hits him in a flash, as Leibniz will later learn. Bouvet is convinced that in the dyadic he has found the key to the meaning of this ancient, no-longer-intelligible system of knowledge. The French priest assigns to the sixty-four hexagrams the binary numbers 0 through 63. He dates the hexagrams back to the time of Noah's children and sees in them a witness to the creation of the world in line with Christian teachings. In this way, Bouvet concludes, one could, by introducing the Chinese to the dyadic, present the biblical story of Genesis to them as lost wisdom from their own tradition.

Bouvet immediately goes to work on his reply to Leibniz. On the

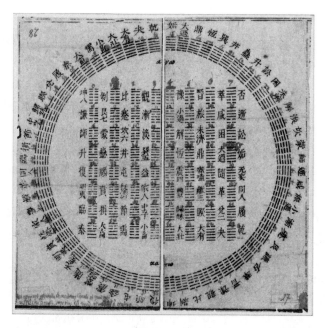

*Joachim Bouvet for Leibniz: Rendering of the sixty-four
hexagrams of the* I Ching *in the Fu Xi sequence
(with marginalia by Leibniz).*

finest silk paper, he writes down his exciting discovery and includes
a rendering of the hexagrams from the time of the Ming dynasty.
He places the hexagrams in two different arrangements, on the
outside in the shape of a circle, on the inside as a square. (Bouvet
also refers in his letter to a magic square.) This circle-square format
will be called the Fu Xi sequence. The letter is dated November
4, 1701.[38] But it will take almost a year and five months for it to
reach its intended recipient. Out of concern that the valuable letter
might fall into the hands of privateers, Bouvet doesn't entrust it to
the French merchant ship *Amphitrite*, as he ordinarily would, but
instead sends it to London onboard an English ship. From there
the letter finally reaches the Jesuit college in Paris and makes it
into the hands of Charles Le Gobien, the procurator for the China
mission. Together with a letter from Jean de Fontenay, Le Gobien
sends the missive from Beijing off to Germany on March 13, 1703.

The intercontinental Jesuit network may not be fast, but it is reliable. And so the parcel containing Bouvet's letter finally reaches Leibniz in Berlin on April 1, 1703.

No sooner is Leibniz, on Brüderstraße, holding the letter from distant China in his hands than it puts him in a state of feverish excitement. Four days later he writes to Father Vota about it in an almost breathless tone ("an astounding thing"; "it is so surprising"; "a wonderful convergence").[39] Three days after that, on April 7, Leibniz writes to Jean-Paul Bignon, director of the royal academies in Paris, and informs him that because of the exciting news from Beijing, he will be reworking and resubmitting the essay on the dyadic that he submitted to the Académie des Sciences some years prior but subsequently withdrew.[40] And finally, on April 17, he turns his gaze to the Royal Society in England. Once his letter reaches Sloane, the two most important scientific societies in Europe will be apprised of the dyadic interpretation of the Chinese oracular text.

MAGIC SQUARE WITH DYADIC SYMBOLS

April, true to its reputation, does whatever it wants. The cold winter lingers; in Berlin the streets have to be cleared of snow and ice several times.[41] The young spring is erratic. In his lodgings on Brüderstraße on this Tuesday, Leibniz is relatively undisturbed. Yet Guidi's letters from Hanover remind him that he can't delay his return forever. The Welf court grows impatient; for too long, Leibniz has neglected his duties there, and by this point Knoche can barely keep his domestic affairs in order. Since late March, Sophie Charlotte hasn't even been in Berlin, having gone to Potsdam by way of Lietzenburg. But Bouvet's enthusiastic letter helps Leibniz tap into unsuspected reserves of energy, plus the pain in his leg is slowly abating. He picks up his quill and explains Bouvet's discovery to Sloane. It is based on two lines, a continuous — for the number

1 and a divided - - for the number 0. So the first four trigrams can be understood as:

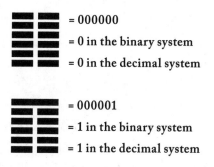

≡≡ = 000 = 0 in the binary system = 0 in the decimal system

≡≡ = 001 = 1 in the binary system = 1 in the decimal system

≡≡ = 010 = 10 in the binary system = 2 in the decimal system

≡≡ = 011 = 11 in the binary system = 3 in the decimal system

And so on. Leibniz draws Sloane a table for the first eight fundamental trigrams of the *I Ching*. This schema is carried forward in the sixty-four hexagrams, for example:

= 000000
= 0 in the binary system
= 0 in the decimal system

= 000001
= 1 in the binary system
= 1 in the decimal system

Leibniz interprets the eight hexagrams in the square's top row, as seen in the sequence that Bouvet sent, as potentially being an account by Emperor Fu Xi of the world's beginning: the 0 can be linked to the void before the creation of heaven and earth. This is followed by the seven days of creation, each represented by a hexagram. The seventh day, the sabbath, is the most perfectly rendered, because 7 in binary form is 111, that is, without a 0; moreover, Leibniz (alluding to Kabbalistic Trinitarianism) sees in the 111 a possible reference to the trinity of scripture.[42] Here Leibniz is abandoning himself completely, letting his mind run wild with Platonic and Pythagorean chains of association. If we want to stick a label on him, we should call him neither the "last" Renaissance magus nor the "first" information scientist.[43] In his case, the two strands, numerology and post-mysticism information theory, seem tightly interwoven. Father Vota catches dyadic fever, too. In his reply on April 17, he speaks of this discovery as at least as significant as the discovery of America.[44] Nor does Leibniz's letter to Sloane on this date fail to produce the desired effect. Its recipient in London has several copies made, which circulate in English scientific circles. And on June 13, the letter is read aloud at a meeting of the Royal Society.[45]

STRUCTURES OF REALITY

We don't know what sort of meals Leibniz partakes of on this day on Brüderstraße. One of them could be—as it often is—kale with beef or chicken, plus a little beer to go with it. It could also be soup. The latter, by his own account, is a mainstay of his diet.[46] Coffee with milk certainly can't have been lacking, and most likely he once again fails to get enough exercise, even with his leg showing improvement. Without a doubt, the seeming correlation between the *I Ching* and the dyadic remains the dominant subject today. Leibniz urges Bouvet to conduct further linguistic research in order to find out the original meaning of the Chinese characters. In language samples from China that he asks Bouvet to send, he is also hoping to find important clues to the history of the spread of Christian peoples, which he'd like to investigate by studying versions of the Lord's Prayer in Asian languages. Eckhart, who could soon be traveling to Moscow, gets roped into this plan as well, also on April 17.[47]

On one crucial point, Bouvet and Leibniz's interests in the hexagrams diverge. In seeking to decode them, the Jesuit priest's primary interest is in reconstructing humanity's sacred first language, which would then enable him to penetrate into the hidden mysteries of creation. Bouvet is convinced that Judeo-Christian messages lie encrypted within the canonical texts of the Chinese. Leibniz's thoughts are focused elsewhere. For him, it's not just the meaning of the characters that matters, but also the relationship between them and the hexagrams. His interest is more formal in nature than anything else. He is looking for clues to the problem of how to connect nonnumerical and purely numerical information. How can ideas be tied to symbols in such a way that the former, like numbers, can be subjected to a formula—a tie that many consider impossible?[48] What is crucial to Leibniz is not the discovery of a hidden reality behind mysterious symbols, but rather the formalization of reality by means of symbols. The idea is to be able to work with these symbols independently of the content they represent, to be able to plug them into a formula and calculate with them in such a way that questions relating to their content can be automatically decided without it being necessary to understand all the mathematical steps taken in between. This is precisely the sense in which Alan Turing will later describe the problem of programming a machine that can translate complex subject matter into symbols, continue processing them, and arrive at a result that one can interpret without having exactly understood the computing operations that form the intermediary steps.

Still, if Leibniz expects to find clues toward a universal language for carrying out such operations, they will come only indirectly. The characters assigned to the hexagrams are composite symbols in their own right, which probably can't be reduced to simple symbols without further ado. Leibniz thus considers the trigrams and hexagrams unsuited to serving as a direct basis for the formulation of a universal symbolic language; at most, they are a passable starting point for the construction of an artificial language for Chinese elites. If, according to Leibniz, any surplus value is to be found here for a universal formula, it is only in the analogous binary structure: the digits 0 and 1 on one hand, the broken and unbroken lines on the other. For him, it is precisely these structural correspondences (also known as isomorphism) that are key.[49] They seem capable of making good on the promise of what a universal artificial language should be able to achieve: rendering various subject areas calculable with one and the same formal system. Leibniz's talent for creative reasoning doesn't consist in delving deep into a single area of knowledge and making new discoveries there; rather, very much in line with his own distinction between field-specific and trans-disciplinary genius, it lies in taking knowledge that was gained in one area (in this case, knowledge concerning an ancient cultural artifact, the *I Ching*) and applying it to a completely different field (here, the development of a universal calculus).[50]

TRANSFER INTO THE FUTURE

Now nothing can keep Leibniz at his desk any longer. He interrupts his reply to Bouvet and eagerly sets out from Brüderstraße to the Royal Prussian Library. It has just occurred to him that he has seen the *I Ching* before, in a Latin translation from 1687. Feverishly he flips through the pages of the library's copy and compares the hexagrams printed in it with those in the illustration Bouvet sent. As he does, he notices significant differences—as he will write to Hans Sloane later today.[51] In the Latin volume, the hexagrams are presented in a different sequence, and he is unable to match the Latin translations of the Chinese characters—for example, air (*caelum*), earth (*terra*), water (*aqua*), mountains (*montes*)—with those on the rendering he received from Beijing. A handwritten Chinese-Spanish dictionary that he consults in the Berlin library doesn't help, either.[52] At the end of the day, Leibniz has to admit that in all likelihood there's still a long way to go before it will be possible to convert knowledge and information into a universal symbolic language and to use the basic symbols to perform calculations in automated, mechanical fashion, governed by a distinct set of algorithmic methods. It won't be so easy, he admits to Bouvet.[53] Leibniz doesn't reach for the stars right away; he's aware that it's going to take a lot of time and a lot of people making an enormous intellectual effort before his bold dreams are made into reality.

On top of that, Leibniz points to the potential limits of such an undertaking. A formal generalization of the kind he has in mind seems at first glance to have potential only for concepts and ideas that are clearly defined, not for those that lack clear definition—here Leibniz is touching on his distinction between clear and confused perceptions. Converting ambiguous ideas into symbols is only partially possible, he says; only those components that have been identified as unambiguous can be converted. In 1931, Kurt Gödel will likewise draw attention to the limits of a logical procedure designed to fully answer all questions.[54] With his work on what will come to be known as Gödel's incompleteness theorems, he will lay the groundwork for theoretical informatics by proving that there is a limit to the statements one can derive from formal systems. A system of this sort functions only if one is able to precisely describe its limits—this is an insight to which Leibniz likewise gives clear expression over the course of these few days in April 1703. And therein lies another specific contribution—largely unrecognized today—that he has made to the long and complex history of the modern digital age. Alongside mathematics, logic, and hardware, a "mosaic tile" for informatics, we may say, also takes shape.

For Leibniz, of course, these loose fragments don't quite seem to fit together. Only in hindsight do they become milestones on the long road from mechanical calculators and bivalent logic to everything from automatic electronic calculators to self-learning robots. But Leibniz should be associated not only with the programmatic idea of a universal method for answering all manner of questions by means of formulas, but also with the suspicion that this might not be possible to an unlimited extent. Leibniz warns not to promise too much, not to promise something that would exceed current human capabilities, even if he also writes to Bouvet that it's important to know what should be done.[55] One can argue over anything, but can one also settle every argument by means of calculation? For Leibniz, the question seems to be not whether this is impossible in principle but whether it's impossible given the current state of human understanding. April 17, 1703, is a prime example of his creative way of thinking in analogies and areas of structural correspondence. In finding such correspondence, he hopes—in line with the idea of an art of invention (*ars inveniendi*)—to bring forth something new: in this case potential new findings for developing a universal method of knowledge and understanding.

Even if Leibniz is only cautiously optimistic in this regard, the parallels between the dyadic and the *I Ching* will continue to fascinate him. Future critical objections to Bouvet's analogy from Louis Bourguet and César Caze will prompt Leibniz to think more about the nature of the correspondence between binary mathematics and the Chinese oracular work. He also holds on to the idea of using the dyadic as an instrument for the Jesuit mission in China.[56] The world of ones and zeroes will captivate him for the rest of his life. Eckhart has a decorative design placed on Leibniz's casket that shows a large 0 with a 1 inside, next to it the words *Omnia ad unum*, "all to one" (in the sense of "everything is connected to one thing").[57] But Leibniz doesn't expect to get a simple formulaic apparatus for universal problem-solving out of binary mathematics. To make the world a little more assessable—that would be enough. To get a bit of support in this confusing world of reason-endowed souls, with all their contradictions. And so Leibniz continues on his search for an "Ariadne's thread" that could lead humankind out of the labyrinth of thought—an image that he uses in his reply to Bouvet. Later Ludwig Wittgenstein will employ a different image: "To show the fly the way out of the fly bottle!"[58]

CHAPTER 5

HANOVER, JANUARY 19, 1710

BETWEEN HISTORY AND NOVEL: HOW GOOD COMES OUT OF EVIL

ALL A MATTER OF PERSPECTIVE

The other's position is the true perspective, as Leibniz once put it, referring to both morality and politics.[1] The idea calls to mind Rosa Luxemburg's famous dictum that freedom is always the freedom of those who think differently. However radical an effort one might make to suspend one's own stance, assuming one's opponent's point of view is certainly illuminating. To what extent Leibniz might have had his own words in mind on January 19, 1710, is unknown. In any case, on this day we see him primarily through the eyes of someone else, the young nobleman Zacharias Konrad von Uffenbach. He and his younger brother Johann Friedrich have traveled from Frankfurt and have come to pay Leibniz a visit. Zacharias Konrad writes of their encounter in his travel diary.[2] It is the third meeting between the three men in the last nine days. This time the two travelers have come to bid Leibniz goodbye. Today, as on the previous two occasions, they won't get what they actually wanted from him—and Leibniz too would rather be doing something different with his time. Instead they talk about this and that. Still, what at first comes off as a polite, if awkward, conversation, nevertheless yields unexpected insights into the world of Leibniz's thought—one need only read between the lines of the older brother's account. In so doing, we catch a glimpse into a fascinating laboratory of associations, a place where many things that seem to have nothing in common are tied together in a surprising way.

On this Sunday, what comes into focus are unexpected connections among history, literature, and philosophy. Most people today know of Leibniz as a philosopher and mathematician; very few know of him as a historian and literary figure. True, it's common knowledge that Leibniz engaged in historical research, but this is usually dismissed as being merely the fulfillment of a duty to his prince—an unpleasant task that only kept him from his actual calling. But what is said—directly or indirectly—on this Sunday in January shows that, to the contrary, Leibniz the historian is passionate about his research. It becomes apparent, moreover, how much history and philosophy are interwoven across the range of his thought, how they enrich and complement his other work. What on the surface comes across as an arid hour of chitchat among three unequal conversation partners turns out, upon closer examination, to be a whirlwind of stories, fables, and fantastic tales concerning freedom, guilt, and necessity.

OUTSIDE THE LIBRARY DOOR

In early November 1709, the two young scions of a venerable old patrician family, following the custom for young members of the upper class at that time, set out from Frankfurt to venture upon an educational journey—a Grand Tour—that will take them all over Germany and northern Europe. After stopping for a while in Hildesheim, Zacharias Konrad and Johann Friedrich von Uffenbach arrive in Hanover in the late afternoon on January 9, 1710. They have a hard time finding a place to stay; it's Carnival time in Hanover, and there are many visitors in the city. Finally they secure lodgings at the Rote Schenke on Calenberger Straße in the Neustadt. Zacharias Konrad is a passionate collector of books, manuscripts, and antique objects of every kind. At twenty-six, he already has amassed an impressive private library; his reputation as a bibliophile precedes the young jurist, who received his doctorate in Halle. Later, as mayor of his hometown, he will assemble one of the most renowned private collections of the eighteenth century, with around forty thousand books and manuscripts.[3] For Zacharias Konrad and his brother, who is four years younger, the main goal of the trip is to visit libraries, collections, and art galleries of note.

On the day after their arrival, a Friday, they want to see the elector's court library and the private library of the "world-renowned and eminently learned Herr geheimden Rath von Leibnitz." At that point in time, Leibniz is neither a Geheimer Rat (he's still "just" a Geheimer Justizrat), nor the bearer of a noble title (*von*). That the brothers Uffenbach refer to him as such demonstrates that Leibniz's work on his own myth is very much bearing fruit. He is certainly famous, but somehow he has also remained an outsider. In the afternoon, the brothers seek out the Lüde House on Schmiedestraße. This richly ornamented Renaissance-era building—which in 1652, at the behest of Secretary of War Carol von Lüde, was renovated and outfitted with a typical Baroque gable roof—was chosen in 1698 by Elector George Louis, son of Ernest Augustus and Sophie, to house not only his princely collection of books but also his enterprising librarian.

Ever since then, Leibniz has lived amid the book collection entrusted to his care. Like his father before him, the elector has had to wrestle with Leibniz's desire for freedom, and he hopes this might be a way to keep him tethered to his duties in Hanover.

The Uffenbachs are shown into Leibniz's rooms. But their actual wish, to be permitted to poke around in the library's collections, goes ungranted. Leibniz refuses their entreaties. Everything is in disarray, they are told. The brothers were warned that Leibniz likes to give everybody the same excuse; it is his way of getting rid of

The Lüde House on Schmiedestraße, with Leibniz's apartments, E. Willmann after G. Osterwald, colorized steel engraving, c. 1860.

visitors. They are denied access to their host's private library as well, with the same argument. What a flimsy pretense, the Uffenbachs think, fuming; Leibniz probably wants to "worm his way" through the books alone.[4] What even Leibniz's contemporaries probably suspect is that he doesn't always make the sharpest distinction between the elector's collections and his own. Whose business is it, he may have thought, if a book or two here and there has some passages underlined or a few notes written in it? But he does bring out a few select volumes from his personal collection for his disappointed guests, and a lively conversation begins. In the evening, after Leibniz has seen them out, the young Uffenbachs plunge into the pleasures of Hanover's Carnival, which this year, like every other, is celebrated for several weeks beginning in early January. Amid the festivities, they try to dispel their disappointment. A raucous party like the masked ball at the town hall is just the thing.

Two days later, on Sunday, January 12, Leibniz looks the travelers up and pays them a return visit at their lodgings, even though Zacharias Konrad von Uffenbach has protested vehemently and would rather be given leave to look through the elector's library. Unfazed by the Uffenbachs' annoyance, the stubborn librarian again draws them into an hours-long conversation. The Uffenbachs put up with it with pained politeness, and over the course of the visit, they learn a lot about Leibniz's current plans to have the Pandects—the digest of Roman legal writings—transcribed word for word, and to rework the catalog of the library in Wolfenbüttel. Leibniz also laments that the English travelers in Greece were interested only in medallions and inscriptions from antiquity and not in ancient codices, and that far more accounts from the age of antiquity and the "middle" age— accounts that bear witness to historical events and other things of note—could be saved from ruin, "and such like."[5] One can sense that the Uffenbachs' heads must have been spinning; meanwhile they've seen only a tiny slice of Leibniz's unparalleled range of activity. He has made sure to talk only about things that he believes will line up with their own interests.

Over the next few days the patricians from Frankfurt explore the city. They seek out private libraries that they have heard or read about, take part in book auctions, and visit a mental calculator, famous in Hanover and beyond, who performs breathtaking feats with numbers. They see the palace and grounds at Herrenhausen and the summer house of General-Erbpostmeister (hereditary postmaster general) Franz Ernst Reichsfreiherr von Platen, who died the previous year. One day they travel out to nearby Loccum, where Abbot Gerhard Wolter Molanus (van der Muelen) shows them the art and natural history collection at the monastery. In the evenings they give themselves over again to the pleasures of Carnival, whether it's attending a masked ball (known as a *Redoute*) at the town hall or the performance of a comedy at the palace opera house.[6] After all, a Baroque nobleman must know how to disport himself, especially when he's on his Grand Tour.

FUR STOCKINGS AND FELT SOCKS

And so late in the afternoon on January 19, the three men sit together one last time. The contrasts could hardly be greater: the two self-assured young noblemen, who have just finished their studies at university and are hungry for success and acclaim, and the ornery scholar, who at "more than sixty years" of age is already an old man and has stalled in his professional ascent, topping out at the position of house historian and court librarian in a provincial North German state. The Uffenbachs have given up all hope of being able to take a peek at the library. Their host, whose "strange appearance" they will later attest to, is probably dressed as he was nine days earlier: in fur stockings, a fur-lined nightgown, large gray felt socks (in lieu of slippers), and an oddly long wig.[7] The giant full-bodied black wig has long since become his trademark. What was once considered a mark of dignity and recognition, however, is slowly going out of style. Without meaning to, Leibniz is increasingly at risk of

*Portrait of
Leibniz ca. 1711
(after Le Bonté,
1788), oil on
canvas (copy).*

coming off as antiquated and grotesque. Even Sophie pokes fun at the many patches in his clothing—they look like rags, she says—and Ernest Augustus, a younger brother of Elector George Louis, compares Leibniz's appearance, with undisguised mockery, to that of the Duke of Wolfenbüttel's court jester.[8]

But the three men also have a lot in common: they love books, manuscripts, and antiques, and they share an interest in politics, law, and history. Like Leibniz, Zacharias Konrad von Uffenbach studied law and is a close acquaintance of the recently deceased Hiob Ludolf, who in the Collegium Imperiale Historicum worked to foster transregional collaboration among German historians. The travelers from Frankfurt appreciate more than anything Leibniz's politeness and talkativeness. In this meeting he comes across as a cheery chatterbox with an enormous store of knowledge, and in a certain respect the book-loving brothers get their money's worth after all, for Leibniz himself is a living library—he is even referred to occasionally as a *viva bibliotheca*.[9] As a librarian, however, his visionary years are behind him. Once he dreamed of a library with two or three rooms that housed a few "core books" in which were recorded all the essential, fundamental concepts from which, in turn, all knowledge could be derived—a universal conceptual framework in miniature, so to speak.[10] Now, however, he has to struggle with repeated cuts to his acquisitions budget and with the wishes of the rulers he serves (since 1691 he has also been head of the Biblioteca Augusta in Wolfenbüttel), who would like him to acquire showy folios above all else. Leibniz is less interested in what looks lordly and distinguished than in what is politically functional.

The prince's library is meant to be part of the information-based exercise of power, a pyramid of political knowledge: at the base are archives and libraries that compile and preserve information; above them are registry agencies and statistics offices where the knowledge is turned into tables of data, which then in turn can be processed in state chancelleries for running the economy, politics, and society. The knowledge thus gained and condensed shall then be placed at the disposal of the prince—at the tip of the knowledge-is-power pyramid—in the form of transportable "state tablets." He should be able to carry these tablets "in a small iron box" so as to be able to look up important data and key concepts at any time to guide his political actions.[11] These fantasies belong to the past. Still, over the years Leibniz has managed, by purchasing several larger collections, to build up the Welf rulers' troves of books and turn them into venerable private libraries. What the Uffenbachs hear in Hanover is that the court library in Leibniz's house, to which they are denied access, is supposed to contain some "fifty thousand books."[12] When Leibniz says they're in disarray, it may be only part pretense. In fact, he has been wrestling for some time now with a rigorous cataloging approach; he keeps coming up with new organizational systems, trying to put the exploding mass of printed matter—in his view, far too many books are being written—in all the different and ever more distinctly defined sciences, in proper order.

THE HISTORIAN'S TOOLS

No sooner have the brothers sat down, probably in the audience chamber, which is located in the oriel on the second floor, than Leibniz draws them into a lively conversation.[13] Once again the talk revolves around manuscripts and publications on topics related to law, politics, and historical scholarship. Leibniz, as becomes clear on this January 19, has more than a superficial interest in history. Despite persistent claims to the contrary, he by no means regards his work on the history of the Welf dynasty as mere drudgery, a job that might ensure him a salary but otherwise distracts him from his true passions, philosophy and mathematics. To this day, Leibniz's efforts as a historian have received little attention from fellow historians, even though his approach to the subject was thoroughly modern. For twenty-five years now, Leibniz has been working on a dynastic history of the House of Welf. But he still hasn't produced anything of substantial length that might appear in book form. Elector George Louis could very much use a public appraisal extolling his ancestral pedigree. Such an appraisal, coming as it would after the recent bestowing of the electoral dignity on the House of Brunswick-Lüneburg, could further solidify George Louis's position as the ninth elector in the Holy Roman Empire; moreover, now that the prospect of his succeeding to the English throne has been held out, it would lend support to his claim. George Louis has taken to referring to the *historia domus* as "the invisible book." Meanwhile Leibniz employs a whole host of secretaries and copyists in his "history workshop" to compile and copy historical sources.[14] In any case, he has no intention of writing a book that merely glamorizes the family history of the House of Welf. He's too interested in even the tiniest of details, and besides, the project has long since ballooned into a wide-ranging history of the Holy Roman Empire in the context of Europe as a whole. More than anything, however, it is Leibniz's thorough, painstaking perfectionism, combined with his critical understanding of the sources, that is responsible for the fact that the history of the Welfs is now threatening to become a balanced scholarly portrayal more than a glorifying panegyric to power.

Time and time again, scholarly accuracy and the demands of objectivity prevent the use of historical arguments for political ends. This is illustrated by a letter that Leibniz is drafting this Sunday. Andreas Gottlieb von Bernstorff, since 1709 the leading minister at the Hanoverian court, had asked Leibniz to look up passages in the *Sachsenspiegel* and *Schwabenspiegel* law books that would support Bernstorff's position in the Mecklenburg estates dispute, which involved the exemption of noble property from taxation. Try as he might, Leibniz can find nothing in the medieval legal codes that would argue either for or against this tax policy. He regrets not being able to help Bernstorff, but his scholarly integrity remains incorruptible, and so to a certain degree he engages in passive resistance against the instrumentalization of history for the sake of political interests.[15]

There's no way of determining whether Leibniz composed this letter before or after his meeting with the Uffenbachs. In any event, large portions of the afternoon's conversation center on his Welf history project. The *Scriptores rerum Brunsvicensium*, a collection of sources edited by Leibniz that is meant to serve as the foundation for the actual historical work, comes up for discussion. The first volume was published in 1707; this afternoon Leibniz and his guests mainly talk about the second volume, which hasn't come out yet. Leibniz isn't always successful in finding the best and most reliable manuscripts. Nevertheless, he is constantly working to identify copying errors, later additions, and other distortions of the documentary record[16]—even as he grapples with the radical skepticism toward historical sources that prevails in his time, an attitude known as historical Pyrrhonism (after the Greek skeptic Pyrrho of Elis). For Leibniz, Pyrrhonic skepticism doesn't involve dispelling all historical certainty; rather, it serves as a starting point for a critical historical scholarship whose task is to separate truth from myth, fact from fiction, even if the effort might not always be completely successful.

Identifying forgeries is part of the work of the historian. When the Uffenbachs speak about Molanus's coin collection, which they had seen when visiting the abbot a few days earlier, Leibniz reports that in Paris a new process has been developed for coating medallions with a special varnish that makes them look like antique artifacts.[17] Leibniz uses the presentation of evidence in a legal context to guide his research as a historian, methodologically distinguishing, like a jurist, between levels of certainty: irrefutable proof, high-probability assumption (presumption), low-probability assumption (conjecture), and finally, mere circumstantial evidence and speculation.[18]

With the elder Uffenbach brother, Leibniz shares an enthusiasm for medieval manuscripts; he is, incidentally, likely one of the first historians to explicitly use the term *middle ages* as a designation for an era.[19] Finally Leibniz gets to talking about various old documents and chronicles. In particular, he focuses in on the various continuations of a chronicle that Martinus Polonus (Martin von Troppau) began in the thirteenth century.[20] The reason for his current interest in the medieval chronicler is made clear in a letter from Bartholomäus Des Bosses that will arrive a few days later. The Jesuit theologian and close confidant of Leibniz informs him that he studied a manuscript of the Polonus chronicle in Hildesheim but was unable to find any reference to the story of Pope Joan.[21] But wait, what's that all about?

A WOMAN ON THE PAPAL THRONE?

From Polonus's work and from other chronicles discussed on this January 19, Leibniz is familiar with the story of "Papissa Johanna," who is said to have lived in the ninth century. Following the pontificate of Leo IV, as the sources would have it, a woman going by the name of Johannes Anglicus occupied the papal office, undetected, for two years and seven months. She had been educated in the sciences in Athens. Called by God to travel to Rome, with her lover in tow, she earned great renown there on account of her extraordinary erudition, all the while disguised as a man. After her election as pope, she became pregnant by her consort. One day, on her way from St. Peter's Basilica to the Lateran Palace, she was surprised by labor pains and, somewhere close to the Colosseum, gave birth to a child. Thenceforth all popes would avoid this site on account of the scandal.[22] In Leibniz's time, the confessional conflicts between Catholics and their opponents color the interpretation of this story. While the former declare it to be pure legend, members of the Protestant and Reformed faiths preach the authenticity of the tale and see in it further evidence of the immorality and disgracefulness of the papal church. Whether the story recounted in the surviving sources is taken to be true or not, both sides are in agreement that the events described are deeply detestable. The story shows the unspeakable acts of which women are sometimes capable.

In Leibniz's hands, however, this negative depiction of women begins to change. With the help of a detailed analysis of the sources, he attempts to prove that the story is just a fable. He repeats neither the usual invective directed toward the female sex nor the putative praise for a male mind inhabiting a female body. Rather, he stresses that it is not uncommon for women to don men's clothing in order to obtain learning and renown. Maybe a woman, disguised as a bishop, had given birth to a child in a crowd of pilgrims, Leibniz speculates (on the lowest level of historical probability, as it were), attempting to identify a possible kernel of truth within the legend. He thus relegates the story to the realm of fable, but as he does so, he dispenses with any misogynistic or confessional slant—an example of applied ecumenism, one might say, of conciliatory broadmindedness in the guise of historical elucidation.

By the time of the Uffenbachs' visit, Leibniz has likely mostly completed his long piece on the *papissa* legend. It is one of the lengthiest stand-alone treatises to come into being as a by-product of the Welf history, or *Annales imperii*. It won't be published, however, until many years later, in 1758, under the florid, melancholy title *Flores sparsi in tumulum papissae* (*Flowers Strewn o'er the Papess's Grave*), no doubt an allusion to the expression commonly used in connection with the historian's work: *florilegium*. One need think only of the *Flores temporum*, a text often ascribed to Polonus that likewise concerns the woman pope.[23] In his own florilegium, written to put the "popess" legend to rest, Leibniz the textual critic deploys the full array of tools at his disposal. The treatise can justifiably be considered an exemplary piece of critical historiography. Nevertheless, to this day it is available only in the original Latin edition.

GOD ON TRIAL

There is no clear record of how Leibniz spent his morning. It's rather unlikely that he went to Sunday worship. Even though his metaphysics has him constantly striving to achieve the unity of faith and reason, among his contemporaries he is not exactly known for being an eager churchgoer; indeed in Hanover his reputation is that of a *Glöwenix*—Plattdeutsch for "nonbeliever," or more literally a "believes-nothing"—because he is only rarely to be found at sermons and prayer services.[24] Without a doubt, though, there must be a pile of letters on his desk, waiting to be answered. Among them are letters from leading members of the Berlin Academy of Sciences, who inform their president that the publication of the planned first issue of the academy's journal, *Miscellanea Berolinensia*, has to be delayed yet again.[25]

Even without the correspondence, plenty of work would still be piled up on Leibniz's desk. When he tells the Uffenbachs that things are disorderly, he's likely referring to more than just the elector's book collection. Still, at least he now keeps an "excerpt cabinet"

in his study for the steadily growing number of papers, large and
small, that he has covered in writing. The cabinet has several rows
of little drawers in which Leibniz has taken to storing the strips and
slips of paper on which he scribbles his notes.[26] Included among the
many papers that surround him are materials for a lengthy book
project that is nearly complete. Its aim is to be a grand synthesis
of philosophy and theology that will contain all his reflections on
metaphysics, from the early, fledgling efforts of his younger years on
up to the *Discourse on Metaphysics* (1686) and the *New System* (1695).
It is to be a fundamental work on the goodness of God, human freedom, and the origin of evil. These essential questions have always
preoccupied him, but only now has he given concrete shape to his
answers and prepared them for publication.

Leibniz received important inspiration for the project a few years
earlier in conversations with the Prussian queen Sophie Charlotte
and her circle. The discussion sessions in Berlin and Lietzenburg
proved fruitful in several respects. Two major works took shape
during this time. In the *Nouveaux essais sur l'entendement humain*
(*New Essays on Human Understanding*), which won't be published
until long after Leibniz's death, he grappled with John Locke's epistemological theses. Parts of the manuscript he wrote on carriage
rides between Hanover and Berlin. A second manuscript, meanwhile, is devoted to refuting Pierre Bayle's critique of Leibniz's program of optimism. Both book projects are closely connected with
the fate of Sophie Charlotte. When Leibniz learned of Locke's
death, then of the death of the young queen—who, when on her
way to Carnival in Hanover, took ill (probably with the flu or a bad
cold) and died, suddenly and unexpectedly, on February 1, 1705—
he stuck the manuscript of the *New Essays*, his principal work on
epistemology, in a drawer.

By contrast, Leibniz continues working on the planned work
refuting Bayle and intends to publish it, not least as a way of honor-
ing the memory of Sophie Charlotte, whose death had plunged him
into a prolonged period of deep grief.[27] By 1707 the manuscript is

largely finished; the Huguenot and native French speaker Charles
Ancillon, currently living in Berlin, has helped refine the style.
But the search for a publisher drags on, until finally Isaac Troyel
in Amsterdam declares himself willing. The work of preparing the
final layout has been ongoing since the fall of last year. Leibniz corrects the proofs arriving in batches from Amsterdam. Around this
time, January 19, he is busy going through the index and writing
an introduction.[28] This is certainly one reason he tries to turn away
visitors like the brothers Uffenbach, rather than lead them on an
hours-long tour of the collections of the library in his home.

Even though the book is almost ready to go to print, it still
doesn't have a title. Leibniz will finally call it *Essais de Théodicée*,
and it will be published later this same year. The book runs to 660
octavo pages, not counting backmatter, and is the only lengthy philosophical work by Leibniz that will be published in his lifetime.[29]
The title reveals the core of the book. The neologism, formed from
combining *theos* (God) and *dike* (justice), implies the question: How
can this world be the best possible when there are so many evils
and flaws in it? How can these imperfections be justified given the
goodness, wisdom, and omnipotence of God? For this is what it all
turns on in the *Theodicy*: God himself is summoned to stand before
the judge's bench of reason and is first accused, then defended,
and finally cleared of all charges by Leibniz, the best of all possible lawyers. Even going back to his earlier writings and letters,
such as the *Discourse on Metaphysics* and the correspondence with
Arnauld, Leibniz has sought to explain that God, in creating the
world, was dependent on the inner compatibility of all components
(what he calls "compossibility"). The world's genesis necessarily
required a limiting of perfection, and any talk of the best possible
also entails a challenge to humanity to participate in the work of
striving toward perfection (see chapter 2). Challenged by Bayle and
others' radical critique, Leibniz sees himself as called on to once
again lay out and defend his understanding of the world, this time
at length. The focus is on what is evil and bad in the world. Physical

evils necessarily result from the finite nature of being; moral evils are possible because individuals have freedom and with their actions can fail to achieve the good. Accusing Bayle of taking a Manichean view, Leibniz rejects such a conception: the evil in the world has no proper existence; it results only indirectly from the absence of good or when something good is withheld from someone. (Leibniz calls this the privation of good.)

FINDING THE POSITIVE IN THE BAD

Against this backdrop, Leibniz musters a whole arsenal of arguments and techniques to justify the existence of evil in the world. Meanwhile the solution to the theodicy problem is already built into the proposition that everything in the world is arranged in orderly and harmonious fashion: for if all things and all processes have a purpose and an inherent goal (*telos*), then—seen from a higher plane—even what seems to be counterproductive must ultimately also have a purpose. The negative must be positive, and unhappiness must be in the service of happiness. What at first seems displeasing often ceases to seem so when one changes one's perspective. He compares this notion with the artistic practice known as anamorphosis, which produces images that reveal their actual content only when seen from a particular angle, in a mirror, or through a special prism. Thus evil is negative only when regarded in isolation; when seen as part of the whole, a wondrous order reveals itself.

A large square, for example, can be divided into many small squares. If one breaks a small piece out of one of these squares, then the remaining figure will be incomplete and imperfect. It shows its complete, perfect beauty only when one reinserts the broken-out piece, which likewise, seen in isolation, appears irregular and ugly. In the realm of morality, the same principle holds: vices are part of the deal; they play a role in bringing forth the greatest possible happiness. King Midas, Leibniz says, felt less rich after he had turned everything he owned into gold. Leibniz praises the manifold nature of existence and imagines how horrid it would be if everything were entirely perfect and brilliant and beautiful: building a library and filling it only with sumptuously bound editions of Virgil; having to sing only arias from Lully's opera *Cadmus and Hermione*; smashing all one's porcelain, that one might drink only out of gold cups; wearing only diamond buttons; eating only partridge; and so on. In music, dissonance is necessary to bring out the harmony. A zither player would be laughed at if he just played the same string over and over again. Everything, seen in the context of the whole, fulfills its proper purpose and contributes to the wealth, fullness, and beauty of the universe. Whatever Leibniz can't teleologize away, he relativizes. When Bayle says that people are evil and unhappy, that there are prisons and poorhouses everywhere, Leibniz replies that the amount of good in human life is incomparably greater than the amount of bad, just as the number of houses is incomparably greater than the number of prisons.

The bad and wicked are necessary to the good and best possible, the *malum* necessary to the *optimum*. As Leibniz's images and analogies—drawn from art, music, moral thought, and geometry—anthropomorphize the whole universe, everything in the world that can be observed and interpreted, in the present and in the past, becomes transcendent, is elevated to the sphere of the divine and thus into a realm held to be inaccessible to human experience and reason. In his role as God's advocate, Leibniz asks a lot of human reason. Don't these things actually belong exclusively in the realm of faith? Is human understanding even capable of making a judgment here? In the courtroom of the theodicy, reason doesn't seem qualified to be the judge; in doing so, it goes beyond its jurisdiction. Immanuel Kant, referring to this problem, will later talk of *Herausvernünfteln*—speculative, extrapolatory reasoning.[30] It's not surprising that Leibniz's apologia for the bad in the world received such a devastating critique in Voltaire's *Candide*. In Voltaire's novel, Dr. Pangloss embodies every cliché of the out-of-touch scholar in his ivory tower, praising his own egotism, war, destruction—in short, the most tragic events, even the horrible Lisbon earthquake of 1755—as all being wonderful and necessary. Scholars debate whether the criticism applies to Leibniz at all and not instead to the proponents of a popular optimism characteristic of the early Enlightenment, an optimism that is nowhere given more succinct expression than in Alexander Pope's poem "An Essay on Man" from 1734, culminating in the line "Whatever is, is right."

Leibniz doesn't see the world naïvely, through rose-colored glasses; rather, he is interested in finding a practical, everyday attitude in the face of potential disaster and the ever-present threat of misfortune. Negative events are unavoidable, but one can learn to deal with them. And Leibniz really can't complain of a lack of setbacks and defeats in his own life. His philosophy grows out of his own experience, out of the knowledge that one can cope with catastrophes, and that one often comes out of crises stronger than before. In fact, he is daily faced with a world in which some people live in great poverty but don't complain, despite the fact that their lot is far from the best; while other people live in wealth and excess but have grown weary of their riches and no longer even realize how fortunate they are. Privileged whining; doubling one's misery by constantly bemoaning it: the paradoxical dialectic of fortune and misfortune is in a certain respect timeless. Leibniz values the contrasts. Scarcity fuels creativity, necessity is the mother of invention. If all one knows is sunshine, it sometimes gets boring; only after rain and gray skies does a sunny day become an especially joyful experience. To wrest something positive out of even unpleasant experiences—to ask, *who knows what good might be found in this?*—helps one get through a lot of things more easily. Leibniz's pragmatic optimism has nothing in common with that of the narrow-minded Dr. Pangloss. Sure, the pessimist might be right in the end, but until then the optimist lives a more contented, less depressed life.

Without a doubt, Voltaire's *Candide* has distorted the view of

Leibniz's disciplined good cheer—but the critique of Voltaire made by certain defenders of Leibniz has distorted the view of the undeniably problematic side of Leibniz's philosophy. What works well on an everyday basis, what lines up with an endearing attitude toward life and its hardships, becomes difficult to sustain when it is so inflated as to become a model for explaining the world, generalized to the point of becoming a cosmological principle. For in that case, all the evil in the world would seem to have some inherently useful or necessary aspect to it. Is it okay, then, to do something bad in order to achieve something good? Does the end justify the means, and does succeeding always make one right? Leibniz certainly isn't out to give people a free pass, morally speaking. But his metaphysics offers few arguments that would prevent such a misunderstanding.

We don't know whether the Uffenbachs were also able to get something positive out of their thwarted attempts to see the library at Leibniz's house. His talkativeness seems to have been some compensation at least. Who knows? If Leibniz had allowed them to take a glimpse past the library's doors, they might not have gotten caught up in such lengthy conversations, and posterity would have been robbed of a rare firsthand account of Baroque discussion culture. From the traveling brothers' point of view, the encounter was not so gratifying, but in the grand scheme of things, maybe it was quite a good thing after all.

WROTH AND BRUTAL, THE LAST IN THE LINE OF THE ROMAN KINGS

So is the time Leibniz spent on historical and literary topics this afternoon no more than an abrupt come-down from the morning's lofty heights of philosophy? Far from it. In the introduction to the *Theodicy*, which he is composing around this time, he writes that since he was a young man, he has been drawn as much to objects of philosophical contemplation as to history and fables. In this spirit, he cites, in addition to Martin Luther's tract countering Erasmus of Rotterdam, the work of Lorenzo (Laurentius) Valla, a humanist writer from the fifteenth century. At the end of the *Theodicy*, Leibniz summarizes the book's theses and arguments in the form of a short story, borrowing an episode from Valla's dialogue on free will (*De libero arbitrio*) and elaborating upon it.[31] The story is centered on a real historical person, the Roman Sextus Tarquinius, eldest son of King Tarquinius Superbus. According to sources from antiquity (Dionysius of Halicarnassus, Livy, and Ovid), the king behaved as a tyrant, committing evil deeds, and was banished by the Romans. Leibniz, like Valla, poses the question of whether the last in the line of the Tarquinian kings could have acted any differently if he had wanted to, or if his fate was inevitable and therefore not he, but ultimately God, should be held responsible for his actions. Valla has Sextus Tarquinius seek out the Oracle at Delphi, who predicts a disastrous future for him. When asked why he couldn't grant him a better fate, Apollo answers that he is not capable of changing the future, for it was fixed by the divine will of Jupiter.[32]

Leibniz continues Valla's story, his own narrative taking the form of several stories nested one inside the other. After Sextus Tarquinius sees Apollo at Delphi, he seeks out Jupiter in the temple at Dodona. There the young Sextus learns that he will commit evil deeds unless he renounces the Roman crown. The head priest at Dodona, Theodorus, observes this scene and, bewildered, appeals to Jupiter, who sends him to his daughter, Pallas Athena. The goddess causes Theodorus to fall asleep in her temple and leads him, in a dream, to the gates of a giant palace. In the Palace of Fates, the head priest glimpses several rooms, in which are represented—as if on a stage—different possible courses that the life of Sextus Tarquinius might follow. One Sextus heeds the warning from the oracle and goes not to Rome but to Corinth, where he dies a wealthy and well-respected man. Another Sextus goes to Thrace, marries the daughter of the otherwise childless king, succeeds to his throne, and is venerated by his subjects. And in the topmost chamber of the palace, Theodorus sees a third, this one the actual Sextus: he leaves the temple of Jupiter seething with rage, travels back to Rome, wreaks havoc there, and violates Lucretia, the wife of his friend. He is subsequently driven off and ends his life unhappily. In every room is a book of fate, and every Sextus bears a number on his forehead that indicates a certain passage in each book in each room. There Theodorus finds the story in all its details. At that moment he awakens from the dream, thanks the goddess, and praises Jupiter's justice. Leibniz believes he has illustrated that divine providence (*providentia*) does not impede the free will of the individual. God does indeed survey the totality of possible worlds but does not limit the individual's freedom, which consists of following the inclination of their unique soul and choosing between a large but ultimately limited number of possibilities.

The short fable at the end of the *Theodicy* shows Leibniz's pronounced fondness for stories and storytelling. Whether his talent in this arena is expressed in poems, brief sketches, or even short plays, it is in high demand at court. Sometimes he even acts, as on one occasion when he portrays the stargazer and astrologer in a play at the Lietzenburg theater. More than anything else, Carnival, which is once again in full swing in Hanover, offers Leibniz repeated opportunities to demonstrate his literary ambitions. Just three years ago he composed a Carnival poem for the Welf court that involved Greek gods disporting themselves in an erotically tinged game.[33] The myths and legends of antiquity are a beloved subject of Leibniz's poetry and prose. In 1702 the "Feast at Trimalchio's" episode from the novel *Satyricon* was performed for a Carnival amusement: Leibniz not only played a minor character but also contributed substantially to the script. The main character of this burlesque comedy is the nouveau riche Trimalchio, a former slave, played by Raugraf Karl Moritz. Trimalchio behaves foolishly and tastelessly at the feast, which is staged at the palace in Hanover as a masked banquet. In courting the lady of his heart's desire, he proves to be—Leibniz wrote the verses for the scene himself—a fiery little Tarquinius, throwing himself at the great Lucretia (or in this case, Hedwig Katharina von Wintzingerode, lady-in-waiting to Electress Sophie).[34] For Leibniz, the poet and philosopher, the ignominious Roman prince who appears in the poetic showdown at the end of the *Theodicy* would seem to be a favorite villain.

TWO FIGURES CAUGHT BETWEEN TRUTH AND FICTION

The record doesn't show whether Leibniz takes part in the Carnival festivities this year. Zacharias Konrad and Johann Friedrich von Uffenbach, however, have a splendid time. The former gives a lengthy account of the celebrations at the town hall: buffet tables with warm drinks and confections; gentlemen dressed up as harlequins, Persians, or peasants; ladies in costume as shepherdesses or Amazons, with long skirts or Venetian masks; the elector himself, almost unrecognizable in his disguise (he doesn't stay long); and the card tables, where "the wench" tries her luck and bets for stakes just as high as those risked by "the cavaliers."[35] While others keep the party going or sleep off their hangovers, Leibniz likely spends this Sunday morning working on the foreword and index of his *Theodicy*. He has to look through the individual entries for concepts and names and send the corrected proofs back to the publisher Troyel in Amsterdam soon. If indeed he spent the morning going through the book, the literary conclusion might be fresh on his mind. In any event, Leibniz doesn't have to leave his philosophy behind when he discusses history and literature with the Uffenbachs in the afternoon. This is evident with regard to two figures who play an important role in his thinking around this time: Pope Joan and Sextus Tarquinius.

At first glance, the two seem to have nothing to do with one another; they appear in different texts that Leibniz just happens to be working on around the same time. The fact that Leibniz is looking for a publisher for both texts has led Des Bosses to assume, incorrectly, that Leibniz wants to have the treatise on the female pope printed as an addendum to the polemic against Bayle.[36] But even though the two works deal with completely different subjects, they have noteworthy connections, and their respective protagonists make clear how closely, for Leibniz, history and philosophy are intertwined. This becomes evident right away at the authorial level: Leibniz not only draws on Valla as the writer of the short passage dealing with Sextus but also mentions him in connection with Pope Joan, holding him up as a model of how to conduct historical criticism, the foremost example being the treatise Valla wrote proving that the Donation of Constantine was a forgery.[37] A spiritual affinity exists between the two scholars: Leibniz appreciates that Valla, like himself, is in his thought and in his writing as much a historian as he is a philosopher.

The parallels continue with regard to the two figures themselves as they feature in the texts. Neither "Papissa" Joan nor Sextus Tarquinius can be fully slotted into either of the opposing categories of historical truth or pure fiction. The last in the line of the Tarquinian kings was almost certainly a real historical person, but the sources contain precious little information, and their reliability is still a matter of debate. Conversely, while Leibniz questions the notion that there once was a female pope, he does consider it possible that a story of cross-dressing—a woman trying to get an education who disguises herself as a bishop—might be the origin of the legend.

There are also striking similarities in the shapes of the narratives. Consider the principal locations where the action in both stories is set: Theodorus goes to Athens to see Pallas; Sextus goes to Rome; the future pope Joan receives her education as a scholar in Athens, then she too travels to Rome. The differences are also noteworthy: the woman disguised as a man follows the gods' advice;

the ambitious prince, by contrast, disregards it. Another conspicuous feature is that Leibniz gives the high priest of Jupiter, who has a positive role in the fable, the name Theodor*us*; but at the same time—unlike the critics of the Catholic Church—he deliberately declines to make any negative comparisons between the *papissa* and Theodor*a*, the legendary Roman prostitute.[38] In both stories, women take on a tragic role. The heavily pregnant woman disguised as a priest is ultimately the victim of a male-dominated world that excludes women from education and science.

Even more dramatic is the fate of the main female character in the Sextus story, Lucretia. Leibniz needs only to hint at it, since everyone—or at least everyone with an educational background similar to his—knows the story already. Let us imagine that the dreaming Theodorus, standing before the book in the topmost chamber of the Palace of Fates, were to turn to the appropriate page. He would be given the following story in detail (according to Dionysius's version): Sextus Tarquinius, visiting his cousin Collatinus at an army camp, is waited on in a friendly manner by Collatinus's wife, Lucretia. That night, captivated by her superior beauty, the king's son attempts to force himself on the object of his desire. She resists him, even when he threatens her with a sword. Only when Sextus declares that he will not only kill her but also murder a slave, lay him naked beside her, and tell her husband he caught the two of them in the shameful act of adultery and punished them at once—only then does she relent to his extortionate demand. The next morning the violated woman, distraught and in tears, hurries to her father and begs him to summon a crowd of people. She tells all those assembled of Sextus's shameful act. As soon as she is finished, she produces a dagger from beneath her robes and stabs herself.

What deserves particular attention in both stories is how they play with different possible lives and destinies. Below the uppermost chamber in the Palace of Fates where Theodorus glimpses the real, historical Sextus, the rooms containing other, merely possible

Sextusses are stacked in the shape of a pyramid whose base continues ever downward and contains an infinite number of chambers.
Just as there is only one world at the top of the pyramid that God
has chosen, there is no possible world that doesn't have other worlds
below it. But the fate of the female pope is also placed within a
potentially manifold framework. Leibniz denies Pope Joan a place
in real history but suggests that a novel be written, based on the
thrilling material of the legend, in which something good comes
out of the story. The *papissa* could die in childbirth without the public learning the cause of her death. Or the mother could survive
along with her male child. He could grow up in the care of friends
and later become an honest citizen or even a heroic, eminent man.[39]
Leibniz seems even to have considered writing a historical novel
based on this story himself. One idea would be to include the saga
of "Joanna papissa" as a subplot in the main narrative of a fantasy
story about Otto the Great and his second wife, Adelaide.[40]

For Leibniz, novels have the status of possible worlds. What holds
true for the merely possible in general also applies to poetry and
literature in particular: literary worlds don't need to become real.
Otherwise one couldn't invent characters for a novel that wouldn't
exist in some future place at some future time.[41] Novels with (more
or less) historical characters are thus nothing other than fictional
history books, like those that are found in the countless rooms of
potential Sextuses in the pyramid of destiny. What's more, historical narratives and invented ones are equally capable of containing
moral messages. Just as the actual Sextus chose an evil fate, so does
Pope Joan, in Leibniz's view, deserve a novelist who will portray
her as a great and courageous heroine. Here Leibniz is thinking of
models from the narrative literature of his time, such as the title
character from John Barclay's *Argenis*, who is saved from a suitor she
doesn't love by a prince disguised as a maiden. Or *Clélie*, a novel by
the famous Parisian author Madeleine de Scudéry, a correspondent
of Leibniz's, whose heroine flees her captors and makes an adven-
turous escape, bravely swimming across the Tiber.[42]

The invented worlds of great literature are subject to the same ordering conditions as the possible worlds straining to exist. The deeper one descends into the pyramid of fate, the greater the degree to which the wealth of possibilities contained in the worlds represented in each of the chambers are internally incompatible. They're too good to be true. No one imitates God better, says Leibniz, than the writers of captivating novels. But reality can sometimes be more surprising than may seem possible. We should take care to note that for Leibniz there is always a sharp line dividing fiction and fact, even if—and this is especially true when it comes to determining the reality of historical events—it's not possible to clearly distinguish between the two. There can be only *one* reality, which for Leibniz rules out not only parallel worlds but also "alternative facts." But literature is just one example of how productive the invented and imagined can be. Even in the "hard" sciences, the fictional can occasionally serve an important purpose. Infinitesimal calculus offers proof that infinite values don't have to be real to be put to use in mathematics, where one can do fantastic things with them (see chapter 1).

The Palace of Fates thus serves as a starting point for exploring the idea of virtual worlds. The Sextus-worlds in the lower chambers may seem deceptively real, but they are not. The only way to determine this, however, is to establish that there is in fact a real world. Without it, each of the merely imagined worlds would be no less real than any other, because there would be no real world from which to distinguish them. Consider the discussion today about the question of whether we might live in a (computer-)simulated world. Is the world a digital decoy, and are we ourselves just electronic circuits? But to speculate in this way assumes that a real world must exist. If something is a simulation, then this implies that there is something being simulated and someone who is doing the simulating. In Leibniz's dream fable, the topmost chamber in the palace is part of the rest of the vision, true, but what is shown within that chamber is meant to correspond to the reality to which Theodorus returns when he wakes. For Leibniz, simple substances, understood in their physical form as perceptible phenomena, guarantee that there is a difference between reality and the merely imagined. Leibniz rejects George Berkeley's conception of things, outlined in a book that the Anglo-Irish philosopher publishes this year, in which he disputes the reality of bodies and considers as real only perception itself (*esse est percipi*).[43] If there were no real world—to extrapolate from Leibniz's argument—then the ontological status of novels would also be unclear. To speak of something being made up only makes sense if something exists that isn't made up.

Crucially, Leibniz also carries the theodicy idea of a (single) real world over into the infinitely many possible and fictional worlds. The actual Sextus who corresponds to the Sextus mentioned in historical sources, chooses evil of his own free will, and God lets this happen because it is part of the best of all possible worlds. For according to Leibniz, great things come out of the Tarquinian prince's crime. Let us once again have Theodorus place his finger on the corresponding lines in the book of fate: all those present cry out in horror when the wife of Collatinus collapses and dies. By Lucretia's bloody dagger, they swear to put an end to the tyranny of the Tarquinian kings. They drive out Sextus and the rest of the royal family, found the Roman Republic, and elect Lucretia's widower as one of the two consuls. Rome is liberated, Leibniz writes, and the seeds of an empire are sown, an empire that will achieve much and "show noble examples to mankind."[44] So runs the teleological interpretation of the events that Theodorus is able to observe in the topmost chamber of the pyramid, which according to the logic of Leibniz's fable align with the actual events in the history of the last Roman king. The *papissa* legend could be rewritten in similar fashion, becoming a novel about a woman who wears men's clothing out of necessity and gives birth to a child who will later contribute wonderful things to the world. Leibniz's theodicy idea, that bad can lead to good, can be applied not only to the historical Sextus (or more accurately, to what was thought to be the historical Sextus) but also to a possible fictional story about a female pope.

THE PROGRESSION OF TIME

Fortunately the Uffenbachs don't arrive until late in the afternoon, so Leibniz doesn't lose too much time today as a result of their visit. By this point, it's dark outside. Of all the phenomena from which we gain our sense of time, the progression between day and night, along with the changing of the seasons, is one of the most important—of this Leibniz is convinced. Clocks, by contrast, only divide the sense of time so gained into regular units. Time and the meaning attached to it are also, ultimately, what make up the crucial link between philosophy and history. Natural phenomena and their basis in astronomy have been instrumental to Leibniz in honing his concept of time, but so have his reflections on history. Here natural science and historical scholarship run together. This is illustrated once again by the central argument he employs to, he believes, finally dispel any claim to truthfulness in the story of the female pope.

For Leibniz, time is not an entity in its own right that exists independently of other things; rather, time exists only in the form of relations between things and their changing states. In this respect, he explicitly sets himself apart from Newton and his followers, who posit the existence of an absolute space and an absolute time. Leibniz, by contrast, defines both as categories of human understanding that make it possible to establish links between causes and effects and to connect disparate events by placing them along a timeline of Before and After. Even a little fly can sometimes be the catalyst for great things. Appearing at the crucial moment, it is capable of distracting a powerful statesman and making his thoughts veer in another direction. Said statesman could, for example, be trying to decide whether to start a war. One thinks of Caesar, halting on the bank of the Rubicon and deliberating over whether he should cross. The fly can stand at the beginning of a new chain of causation. To observe this means to put the individual links in this chain in chronological order, just as Theodorus looks on and sees that Sextus reaches the disastrous decision to travel to Rome only after having left the temple of Jupiter in a fury.

Leibniz defines space as the order of things that exist simultaneously, and time as the order of things or states that exist in succession. To say that God could have created the world a year earlier wouldn't make any sense if one had only an absolute notion of time rather than two points in time between which one could draw a relation. Time can also be depicted spatially, as in genealogy: in the drawing of a family tree, the degrees of kinship between family members are drawn with short or long lines, with or without other relations in between. For Leibniz, time can move in only one direction; there is no reverse causality. Zacharias Konrad von Uffenbach's account of his visit to Leibniz's home also follows a single chronological progression. And it is precisely this notion of events occurring in sequence, one after the other, that consigns Pope Joan to the realm of legend. After an exhaustive analysis of the relevant sources, Leibniz establishes a chronology indicating an unbroken succession of reigning popes in which there is no room for the *papissa*. Those who argue that Pope Joan existed have gotten the math wrong, Leibniz says; they have placed the death of Pope Leo IV a year too early (in 854) and the death of Pope Benedict III a year too late (in 859). According to Leibniz's chronology, the three popes Leo IV, Benedict III, and Nicholas I ruled in immediate succession.

It is that order of time that torpedoes the fable of the female pope. "*Explosa Papissae fabula*," Leibniz writes, sober and restrained, expressly declining to offer the story up as any sort of antichurch polemic (Rome as the whore of Babylon) or defamatory, misogynistic screed (Pope Joan as female monster).[45] In the courtroom of historical criticism, the scholar of history hands down his sentence: the *papissa* is banned from the past. It isn't God's justice that has to defend itself here, it's time—"chronodicy," one might call it.[46] Sextus Tarquinius, Pope Joan, Theodorus, and Lucretia—in the knotlike tangle of the world of Leibnizian thought, people and things are often linked together in rather striking and unexpected ways:

one enters the labyrinth through the bottom left exit, labeled *Theodicy*, and comes back out again through the entrance at the top right, where one finds the history of the Welfs. Or vice versa.

The two Uffenbach brothers, if nothing else, don't seem to have been bored this afternoon. Books open their readers up to new worlds—on this all three are in agreement. They encourage readers to think in terms of possibilities. This comes more easily to optimists—of this, Leibniz is certain. Allowing different possibilities to reveal themselves always means shifting and reconfiguring one's own perspective, which also makes it easier to see the positive. . . . The brothers' heads must have been humming from sheer overload of new impressions and information. After they depart from Leibniz's company, they once again seek diversion and entertainment in Carnival—that evening a "merry comedy" is being performed at the Welfenschloss. Leibniz, for his part, likely stays home this evening, where he probably partakes, as he has been doing more frequently of late, of "a sturdy repast," after which he immediately goes to bed.[47] The two noblemen from Frankfurt will leave Hanover the next day and continue on their journey. Later it will occur to the elder Uffenbach that there was still so much he'd wanted to ask Leibniz, but he unfortunately forgot until after they parted. Maybe Leibniz should have told him more about his interpretation of the *papissa* legend. Maybe then Zacharias Konrad would have reconsidered his opinion that the story was true and a shameful stain on the Catholic Church. Maybe then, later on, after speaking of "Pope Joan" with a friend, he would have ended their discussion differently, and not with the words "But that's enough of the old whore."[48]

CHAPTER 6

VIENNA, AUGUST 26, 1714

INTERCONNECTED ISOLATION: TORN BETWEEN SOLITUDE AND TOGETHERNESS

DEEP BREATH . . . AND ANOTHER

The news comes hard and fast. What a weekend! Yesterday he took care of the most important mail; today he has a moment of calm for a little philosophy before diving back into the onrushing flood of events. Leibniz is getting ready to leave Vienna. But he's still waiting for word from Hanover. When should he head out? He hasn't taken care of everything he needs to yet, not even close. The last few months have been very productive. Many a metaphysical meditation has been set down on paper. Leibniz keeps tinkering with them; he simply can't let go, crossing words out here and adding words there, on and on. But now there's no time left. He has to take all his Viennese papers back to Hanover with him. When he will find the time to resume work on them is for the stars to decide, but it won't be in the next few weeks.

What he holds in his hands, on this late summer Sunday in Vienna, are notes and sketches for two works that today are considered key late contributions to his philosophical oeuvre. Like many of Leibniz's writings, they begin as clarifications or comments made in response to questions or critical objections. They seem almost like random snippets from the incessant churn of thoughts whirling inside of his head. Again and again the same questions arise concerning God, existence, science, freedom, and morality—and the big question of how it all fits together. As Leibniz reaches for his quill and again puts ink to paper, the perspective shifts, imperceptibly, a tiny little bit each time. Those who know me only from my printed writings don't know me at all, Leibniz is often quoted as saying.[1] By this point he is sixty-eight years old, a formidable age for that time. But still his quill will not rest. And so, on this August Sunday, there are more newly begun writings on his desk in Vienna, destined for revision or for the copyist, who will make a fair copy of them. What he sets down on paper in these uneasy summer weeks displays—to a more extreme extent than in previous years—a tension between, on the one hand, the individuality of being, which Leibniz describes as impenetrable and self-contained, and on the other the many-sided interconnectedness of all being. Both elements seem to be growing ever less compatible, not only in Leibniz's thought but in his life as well.

SUCCESS AND SOLITUDE

Only when he pauses for a brief moment does it dawn on Leibniz just how much has happened in the past few weeks and months. He's been living in Vienna for more than twenty months now. On this weekend in August, he looks back on all that's happened in that time. As he does so, many things come to light that reveal the increasingly divided nature of his situation. On one level, it would make sense for him to feel that he has achieved all he wanted. By this point, he's got titles and offices galore. He is a Geheimer Justizrat to the Elector of Hanover, the King of Prussia, and as of November 1712, to the Russian czar Peter I as well. He is a member of scholarly societies in Paris and London and president of the

Berlin Academy of Sciences, on top of which he is still the court librarian in Wolfenbüttel and Hanover, and still the historian of the House of Welf. But that's not all. Following a meeting with the czar in Karlsbad (Karlovy Vary) in December 1712, Leibniz travels on to Vienna for what is at least his seventh trip to the capital of the Holy Roman Empire since 1688. After Leopold I and Joseph I, Charles VI becomes the third emperor to whom he has offered his services. Not only does he become an imperial adviser, but he also receives the title of Reichshofrat (imperial court councilor).[2] Since then, he has enjoyed the privilege of private audiences with the emperor, has met with many ministers and state officials, and has been vying for the position of chancellor of Transylvania. With a combined salary of around 8,000 guilders, he commands a nominal annual income higher than that of the highest-paid official at the Hanoverian court.

At the same time, however, Leibniz finds himself increasingly isolated and boxed out. The bulk of his salary remains nothing but paper and promises. He waits in vain for the money to be paid out. Time and time again he is given reassurances and is forced to appeal directly to the emperor and remind him of the outstanding payments, most recently at the start of last month, July 1714. Meanwhile his estrangement from the electoral court in Hanover is growing. Leibniz had traveled to Vienna without the permission of George Louis; at first the Hanoverian court believed he would be returning soon, but in the fall of the previous year, the electoral administration had stopped paying his salary. Vienna is an expensive city, and it's only thanks to his accumulated savings that Leibniz is able to keep his head above water. His mind is ever more preoccupied with the worry of becoming impoverished in his old age.

The outlook isn't any better when he turns his gaze toward Berlin. Being so far away, he is scarcely able to exert any influence on the direction of the academy he founded. Since his last stay on the Spree in 1711, the activities of the academy have left something to be desired. In London, moreover, dark clouds continue to gather in the priority debate over differential and integral calculus. With the publication of the *Commercium epistolicum*, a short tract that has been circulating among scholars all over Europe, the Royal Society has officially accused Leibniz of plagiarism and declared that Newton was the first to come up with infinitesimal calculus.[3] Leibniz is infuriated. European mathematicians have split into two camps, Leibniz supporters and Leibniz critics.

In Vienna, Leibniz lives on the Lugeck in the Großer Federlhof, a stately Renaissance building with a six-story tower and courtyard arcades near St. Stephan's Cathedral. Despite his privileged lodgings in the heart of the city, gaining access to Vienna's center of power remains a constant struggle. This felt especially true last year, when the plague was raging in Vienna. Larger and larger swaths of the city had been gripped by the deadly disease. The epidemic drew dangerously close, and by October it reached his door. Leibniz had to leave his lodgings after seeing with his own eyes a mortally ill man being carried out of a neighboring building. Oh, but that was far away, on the other side of the street, he wrote to Charlotte Elisabeth von Klencke, lady-in-waiting to Dowager Empress Wilhelmine Amalie, downplaying the incident.[4]

More than the dangerous disease, he feared no longer being admitted into the emperor's chambers on account of the acute risk of infection. And with good reason, too, since only those who could prove they came from an untouched household were received at the Hofburg. Even high-ranking dignitaries, if they were admitted, were allowed to have only one page or footman accompany them.[5]

THE DISTANT PROXIMITY TO POWER

Since the start of 1714, Leibniz has been back in his rooms in the Großer Federlhof. The number of plague deaths is steadily decreasing. Finally in March the restrictive measures put in place to contain the epidemic are lifted. Taverns open back up; life returns to the streets. Emperor Charles VI sets about fulfilling his solemn oath, taken last fall, to build a church, after the plague had been overcome, dedicated to Saint Charles Borromeo, revered for tending to plague victims. Now that the scourge and the restrictions on movement are in the past, Vienna breathes a sigh of relief. But Leibniz's uncertainty as to how close he actually is to the court remains.

Often he shares a table with major political players, dining with them and joining in their discussions. Among them is the powerful field commander and statesman Prince Eugene of Savoy. He has made a name for himself in the conflict against the Ottomans and in the War of the Spanish Succession, but he also cultivates an interest in science, art, and philosophy. On several occasions this spring and summer, following Eugene's return to Vienna in March from the peace negotiations at Rastatt between the empire and France, Leibniz is a guest at the prince's city palace. Again and again Leibniz tries to wield political influence: in the arena of religious politics, through attempts to lend fresh momentum to the reunification talks between Catholics and Protestants; or the field of imperial martial policy, by advocating for an alliance between Austria and Russia against the Bourbons or by trying to talk Charles VI into funding privateers to harass French ships in the waters off the West Indies.

Leibniz's militaristic side comes out often in Vienna. As a practically minded court scholar of the bellicose Baroque Age, he is conversant in military matters as a matter of course. Strategy, fortification building, and weapons technology have from the beginning featured in his political projects for the powerful. Once he even tried his hand at building a rapid-firing rifle. But in the summer of 1714, he's primarily interested in the logistics of financing large-scale armies, equipping an imperial armada, and provisioning troops and providing them with weather-resistant, durable clothing.[6] All told, Leibniz looses a torrent of proposals and reform ideas upon the government in Vienna in the form of position papers and petitions. Once again he lays out his complete program for societal optimization: restructuring the state administration, reorganizing the police force and judicial system, instituting school reform, and above all founding a scientific academy—not to mention, of course, economic initiatives like establishing a state bank, plus taking measures to contain flooding along the Danube. Nor could he fail to include suggestions for improving public health. Leibniz observed that people sick with the plague and those only suspected of being infected were crammed together in close quarters, either in makeshift hospitals or in small quarantine huts on islands in the Danube; he now proposes building barracks spaced far apart from one another so as to prevent the spread of contagious diseases.[7]

Leibniz's flood of ideas never dries up. Coming up with new ones is what keeps him alive. But he is also aware of his advanced age. He knows that at this point. most of the things he envisions he won't live to see realized. It's enough for him to serve the future of humanity. As a result, it doesn't bother him too much when, as so often happens, many of his ideas come to nothing. The plans to establish a Viennese Academy of Sciences, for example, founder for lack of funds. And it's always an open question whether his political schemes even reach the ears of the powerful. Leibniz tends on occasion to overestimate his influence on high politics. But at bottom he knows that, as a scholar, he doesn't have the status of a professional politician or diplomat. He is merely sitting in the antechamber of power—and he is reminded of this again and again as these summer months go by.

STRAINED LONG-DISTANCE RELATIONSHIPS

The Vienna of Leibniz's time is not yet the fully established player in European politics that it will become under later Hapsburg monarchs. True, the city on the Danube does seem to have decisively shaken off the threat posed by the Turks, but the Ottomans haven't yet been driven out of Hungary or the Balkans, not conclusively. Both the court and the city are asserting themselves, displaying an increased need to project power in the Baroque manner. This year, for example, Prince Eugene commissions the construction of a stately summer palace outside Vienna's walls, today's Belvedere. Rivalries among the great powers of Europe don't play out just on the battlefield but carry over into the symbolic arenas of art, court festivities, patronage, architecture, and music. Still, while the Baroque topography of churches and palaces dominates the look of the city, in its winding alleyways the emancipation of the burgher class has begun, even if the day when this citizenry can aptly be caricatured as the "cozy Vienna" of the imperial-royal monarchy, with its coffeehouses and opera culture, is a still long way off.[8]

How uncozy Leibniz's position is at the moment is evident from the way his relations with the court in Hanover have worsened over the past months. Even after his salary is cut off at the source, Leibniz simply cannot be persuaded to return to the city on the Leine. Time and again he puts off his direct superior, Premierminister Andreas Gottlieb von Bernstorff. Over the winter, there was grumbling in Hanover about Leibniz preferring the plague-ridden air of Vienna to the fresh breezes of northern Germany. Meanwhile Leibniz's servant Johann Friedrich Hodann looks after the deserted house on Schmiedestraße. Once or twice a week he writes his employer in Vienna, forwarding him the mail and books piling up in Hanover in his absence. Hodann complains of being short on money; important things need taking care of, but he can't attend to them because he lacks authority; and as he persistently reminds Leibniz, people keep asking him when his boss will finally be coming back.[9]

Does Leibniz even want to return to Hanover, or would he rather stay in Vienna permanently? No one knows exactly, and one gets the impression that he doesn't know himself. He'd rather keep all his options open. He plays for time, keeps putting off making a decision. Leibniz is left even more adrift when, in the spring, his two biggest patrons and supporters die: Duke Anthony Ulrich on March 8, at the age of eighty, and Electress Sophie on June 8, at eighty-three. The best-case scenario would be to stay in Vienna and at the same time continue working for the Welf courts in Hanover and Wolfenbüttel. Leibniz would like to say yes to several things at once; he's never really learned to say no. One minute his mind is in one place, the next it's somewhere else. Looking past the network of his relationships, he doesn't seem to really be present anywhere; in the thicket of his tangled connections, he is impossible to pin down. Vienna only seems to make the situation worse. Early on the Hanoverian envoy Daniel Erasmus von Huldenberg warned the emperor and his widowed mother: Leibniz wanted to do everything at once; if he had his way, he would be on the move all the time, but he lacked either the talent or the inclination to see anything through to completion.[10]

The government in Hanover seems to have run out of patience. Johann Georg Eckhart informs Leibniz on June 24 that he happened to run into Premierminister Bernstorff on the way home from church. Bernstorff had asked him to present Leibniz with a final ultimatum: he must now decide as quickly as possible whether he will return to Hanover.[11] The ultimatum fails to make an impression on Leibniz. He offers verbose but less-than-convincing justifications, believing he can hold off his superiors on the Leine for a little longer. Yet again there is little movement on the matter, and so the days keep drifting along. . . .

It's been an unusually hot summer, Leibniz learns from Hodann, everyone's afraid of inflation, and in the nearby village of Springe twenty-four buildings burned down. Thankfully the city of Hanover has been spared from the plague. Church services have been held to give thanks for the danger averted, and there has even been a festival of thanksgiving.[12] All quiet on the northern front, then, for the most part. On August 15, Leibniz informs his employee that he definitely intends to return to Hanover before the winter. Who's going to believe him now? Again and again Leibniz remains hazy and vague. To the ears of Hodann, Bernstorff, and others, it must sound like just another delaying tactic. But then a piece of news arrives in Vienna that changes everything in an instant.

TURBULENT DAYS

Hard to believe—it's been only a few days since the news from London and Hanover reached Vienna and turned the capital's political life upside down. Leibniz, looking back on it all on August 26, can still remember every little detail. Just a week earlier, on August 20, a special messenger arrived from Hanover and reported that in England, on August 11—"three hours after noon," as Leibniz hears it—Queen Anne collapsed after several grueling parliamentary sessions.[13] When the news breaks, Leibniz is in the middle of a conversation with Prince Eugene. The prince has informed him that he will be traveling to Baden in Aargau to clear up important political questions between the empire and France that were left open following the conclusion of the Peace of Rastatt. That afternoon Leibniz, along with Hofkanzler Philipp Ludwig Wenzel Count von Sinzendorf, goes to see the Dowager Empress Amalie at Schönbrunn Palace to discuss the news from London and Hanover.

That news keeps coming. The very next day, August 21, another messenger arrives from Hanover. Now it's certain: Queen Anne died on August 12 (August 1 by the Julian calendar). The queen is dead, long live the king! According to the order of succession established by the Act of Settlement of 1701, the crown now passes to the protestant House of Welf in Hanover. There is of course a strong Catholic opposition in London, led by the politician Henry St. John Bolingbroke, but events have clearly taken them by surprise. That very same day Elector George Louis is declared King George I; a courier is sent to Hanover at once to invite the new king to England. Looking back on Saturday, August 25, Leibniz tries to think when the news must have arrived in Hanover. Not before August 16, he figures, because Hodann's letter from this date makes no mention of it.[14] Leibniz is correct. A few days later he receives a vivid report from Hodann: After the queen died, at eight o'clock in the morning on August 12, a force nearly forty thousand strong had gathered outside of the Houses of Parliament in London to prevent any potential unrest. Parliament had confirmed the choice of the new head of state in a unanimous vote, and the courier was dispatched to Elector George Louis immediately thereafter. Hodann goes on: "The courier arrived in Hanover sometime in the night of August 16/17, the envoy Count Clarendon went with him out to Herrenhausen, roused the elector from his sleep around twelve o'clock, and, kneeling, gave him the news of the queen's death. People are saying that the order was given to name the elector His Royal Majesty on August 17. Apparently he was declared king under the name Georgii the First and the proclamation was read out to the accompaniment of trumpet fanfares in various places in London."[15] A historic moment—reading Hodann's lines, the drama of these hours is tangible.

In his mind, Leibniz again runs through all that's happened: last Tuesday, when the second messenger from Hanover arrived, Leibniz had a long conversation with Count Sinzendorf. The following day, August 22, at a dinner with high-ranking Austrian diplomats, he learned that Prince Eugene would probably be delaying his trip to Baden till the following Monday. Leibniz hastens to report all this to Bernstorff that same day, bursting with pride at having the privilege of experiencing these days of upheaval in the company of the Viennese political elite. He must also be sure to let his political contacts in Berlin hear from him right away: that same Wednesday he writes in all haste to Matthias Johann von der Schulenburg, a former lieutenant general who fought for Saxony in the War of the Spanish Succession—but it's too late to make a fair copy of the draft letter before the mail goes out.[16]

Just like that, the situation for Leibniz has changed. He is given unmistakably clear orders to leave Vienna and get himself back to Hanover. This suits him just fine, however, seeing that he hopes to be able to go to London as part of George I's retinue and there perhaps take over the office of court historian. At the same time, he sees an opportunity to help shape major political developments.

Closest to his heart is the idea of improving relations between the empire and England, which in signing the Peace of Utrecht had pulled out of the alliance against France. Leibniz can see it taking shape in his mind's eye: a grand alliance between the Welf king in London, the Electorate of Hanover, and the imperial court in Vienna. If he could be certain that he would still find George I in Hanover, he would set off immediately with the mounted post, Leibniz informs Hodann on Saturday. Full of impatience, he waits for a third messenger, longing "daily, indeed hourly," for him to come so that he can learn when the king and his retinue will be departing for England.[17] Meanwhile the entire Hanoverian court is in a flurry of excitement. Everything is helter-skelter. Who should go with the king to London, who will follow later on, who will stay behind in Hanover? Amid this chaos Johann Georg Eckhart writes breathlessly: "There's so much confusion here among everybody there's no describing it."[18]

Even without this report, which Leibniz won't receive until a few days from now, he can imagine on this August 26 that it must be a madhouse up there on the Leine. In Hanover, too, changes are afoot. Just the day before, on Saturday, Eckhart was named historiographer.[19] In Vienna, the unsuspecting Leibniz doesn't yet know that his own colleague has become his closest rival, or that Eckhart is now gunning to take over the ongoing Welf history project. That same day the Scottish diplomat John Ker of Kersland, who is witnessing firsthand the frantic preparations to leave, admonishes Leibniz to "make haste to Hanover." The necessary political expertise is sorely lacking here, Ker writes; the Hanoverian ministers are "altogether unacquainted" with English affairs. Even Bernstorff, hopelessly overwhelmed by the intrigues between the Whigs and the Tories, "is led by the nose in those matters" by an ignorant English lord.[20]

Everyone, it seems, is eager for departure: Hanover's political leadership for the Thames, and Prince Eugene and his diplomats for Switzerland. But Leibniz sits in Vienna, paralyzed with indecision.

Should he head north right away, or would it be better to wait for the next courier from Hanover, then eventually join the rest of the court before they cross the Channel into England? And one more urgent matter is pressing on him. He has promised Prince Eugene that he will give him a book with a selection of his own writings, so that the prince might better understand his philosophy. And so he has an anthology put together that contains excerpts copied out from older treatises and new meditations and letters composed in Vienna, among them a text written especially for Prince Eugene that bears the title *Principes de la nature et de la grâce fondés en raison* (*Principles of Nature and of Grace, Founded on Reason*). The texts, six of them in all and written in French, are bound in red Morocco leather and emblazoned with the prince's escutcheon in gold. He has to hurry if he wants to present the handsome handwritten volume to the great statesman in person before his departure. The last piece hasn't been copied yet, and there's no time left to get a copyist, so Leibniz adds the final lines to the volume himself. It's possible he hands the—later famous—anthology to the prince in person when he meets with him one last time between Saturday and Monday.[21] Thankfully he has kept copies of all the texts. When he makes ready for his departure from Vienna the following day, he will stash them away in the luggage he means to take with him to Hanover.

SUITCASES PACKED

After days of excitement, a moment of calm sets in on August 26. This Sunday even the court grants itself a rest from the strenuous business of government. Charles VI, the dowager empress, and the closest members of their entourage attend a mass at the Convent of St. Clara in the city center, remaining afterward to witness the solemn initiation (*investiture*) of two nuns. Aside from this, the *Wienerisches Diarium*, a news sheet published twice a week, has nothing to report on public life in the city that day.[22] The great bell known as the Pummerin rings out from St. Stephan's Cathedral—it was cast using metal from melted-down Turkish cannons after the siege of 1683—while not far away, at the Großer Federlhof, Leibniz reads through the letters he wrote that morning. Among them is a letter to John Chamberlayne. As a member of the Royal Society, Chamberlayne had been working to bring about a reconciliation between Leibniz and Newton in the priority dispute. Leibniz writes him to say that as soon as he gets back to Hanover, he will write a paper in his own defense that will include unpublished documents. Newton will later read Leibniz's letter to Chamberlayne with great interest and make a copy of it himself.[23]

The time has come to pack. A whole stack of manuscripts and letters sits ready to go; it all has to be stowed away in his luggage. Along with the copies of texts from the anthology presented to Prince Eugene, the stack includes several pages of notes on folio-size draft paper and foolscap that are intended for Nicolas-François Rémond (also spelled Remond). Rémond, eight years older than Leibniz, is the leading adviser to the Duke of Orléans, an influential aristocrat who regularly brings together an exclusive group of philosophically and literarily inclined poets, scholars, and politicians at his salon in Paris. For some time now, Leibniz's philosophy has been a dominant topic of discussion among the illustrious company. There is much admiration for the German polymath's *Theodicy*, although it seems to raise more questions than it answers. In April 1714, Leibniz had learned through Charles Hugony that the sensitive Rémond was struggling to understand Leibniz's demanding meditations, and his delicate health was suffering as a result; consequently, Hugony informs him, the Rémond circle would like to receive further clarification of the concept of monads as quickly as possible.[24] The explanation of monads (*Éclairissement sur les Monades*) that he begins upon receiving this request—around the same time that he writes the *Principles*—is not yet finished when Leibniz picks up his quill on this Sunday to write to Rémond. What is missing in particular are the references he plans to make to relevant passages in

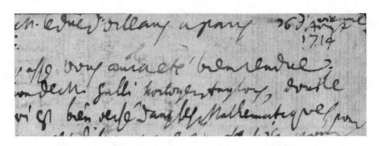

Leibniz to Nicolas-François Rémond, August 26, 1714,
extract in Leibniz's hand. The date and location are visible
in the top right corner: "Vienne 26 d'Aoust."

the *Theodicy.* Leibniz is convinced that he won't have a chance in the foreseeable future to add them and finish the text, which will later become famous under the title *The Monadology.*[25]

So instead Leibniz today is making a copy of the *Principles* to send to Rémond, despite having dedicated it to Prince Eugene. The English clockmaker Henry Sully will be delivering it. May this little discourse (*petit discours*) make his meditations more comprehensible, he writes in an accompanying letter. Truth is more widespread than is commonly assumed, Leibniz assures Rémond. In speaking to the scholastics, he has expressed himself in the language of the Aristotelians, while in speaking to the Cartesians he has adopted their terminology. And with regard to the thinkers of antiquity—Leibniz alludes here to Rémond's affinity for Plato—there was gold to be dug from that ground as well, and diamonds to be found in those mines.

Leibniz doesn't cast the ancient philosophers aside; rather, he sees himself alongside them in an ongoing tradition of *philosophia perennis,* in accordance with which certain insights and pieces of wisdom endure and are perpetuated throughout different eras and across different cultures.[26] And so while the political Leibniz looks mainly toward England and Hanover, the philosophical Leibniz looks toward France and sees himself in a dialogue with the great figures of intellectual history.

LOVE AND GEOMETRY

One thing is certain: Leibniz will miss the selection of good Viennese coffee. As he will miss the sweet Hungarian wine that hardly needs any sugar added to it. And how much he enjoyed stopping in at nearby taverns to eat or take food home with him: the elegant Geitzerhof and Steyerhof, for example, but also taverns like the Drey Haasen and the Schwartzer Elephant that cater to a burgher clientele. At the weekly markets, he could stock up on lemons, pomegranates, sardines, chestnuts, and candied fruits, as well as coffee and chocolate, of course.[27] Leibniz may be feeling a touch of melancholy as he begins to pack up the *Principles* and the *Monadology*, along with many other pieces of writing.

Some texts are already far along; others exist only as vague, preliminary notes. As they so often do, Leibniz's thoughts move in disparate directions, but they come together in the end—usually at some unexpected point. One example is the manuscript of a lecture that he gave to a small, casual gathering (who was present is unknown) on July 1.[28] The stated topic is the philosophy of the ancient Greeks and its contribution to natural and revealed theology. But then Leibniz turns to questions of ethics and justice, taking on the subject of true, selfless love. He distinguishes this love both from egocentric love of self and from the kind of love that follows no self-interest whatsoever but rather is so radically merged with the love of God that individual identity utterly evaporates. (Here Leibniz is thinking of the quietism of François Fénelon, among other examples.)

In reflections such as these, Leibniz is implicitly building on what he describes, in the concluding section of his explanation on the concept of the monad for Rémond, as the community of minds in a morally perfect world. In order to understand this concept, one must know what Leibniz means by true love. For him—and he has stressed this again and again, beginning with his *Confessio philosophi* of 1673—true love is nothing other than every individual's love for their neighbor and desire for their neighbor's happiness. If this wish could be realized on a comprehensive scale, it alone would be sufficient to ensure that the world reached a state of moral perfection. This is because—and here Leibniz's notion of substance comes into play—in the world of monads, each monad does nothing but express and reflect every other monad. And if this reciprocal reflection were driven solely by the wish of each to make the other happy, then the world would automatically be pervaded with love. Line up Leibniz's writings from this spring and summer, and it becomes clear: in Leibniz's thought, justice, morality, and metaphysics form an inseparable, coherent whole.

We see further examples of this in the thoughts that preoccupy him in these late August days leading up to his departure from Vienna. It's possible he hasn't even started to put them in writing yet. Maybe they won't be set down on paper until his upcoming journey to Hanover. The state of the surviving manuscript, very difficult to read and possibly written with his left hand, suggests that he could have worked on it en route while sitting in the coach. The untitled notes begin with the words *Spatium absolutum* (absolute space).[29] Here Leibniz returns to a mathematical subject that he has wrestled with since his Paris years. It is the foundation of a new kind of geometry that is concerned with the spatial relationship between geometric figures or objects. The notion determining the reciprocal spatial relation between them Leibniz calls *situs* (position or situation); the systematic study of them within the framework of a new symbolic formalism is, in turn, *analysis situs*. His new situational geometry, which long remained buried among his manuscripts, anticipates important insights that nineteenth-century mathematical topology will arrive at. Leibniz continues to stick with the Euclidean concept of a three-dimensional space (length, depth, and height), though there are still unsolved problems of classical geometry that he doesn't succeed in clearing up. He can imagine elliptical or warped spaces only as hypotheticals; he doesn't believe they exist.

But the very fact that he is unsuccessful in resolving certain geometrical questions constitutes an important step in the direction of later non-Euclidean geometry.[30]

From this vantage, Leibniz can now sharpen his notion of a relational space, directed against Newton's concept of an absolute space. According to Leibniz, space has no reality of its own separate from things. Rather, it exists only as a system of relations between things or elements (and not as their container). Absolute space is therefore nothing but the sum of all relations between things. On its own, an element is only an abstract point; it begins to be distinct from others only once it forms a positional relation to them. With his new geometry, Leibniz hopes to be able to analyze these positional relations in an exact manner. It follows that space can be defined as a structural system of geometrically describable relation-networks between points. Mathematical points, however, are ideal elements; in the real world, they correspond to monads, which are likewise dimensionless ("without extension") and cannot be conceived of as spatial entities. Yet monads can certainly be understood as physical phenomena, and indeed—to draw an analogy to Leibniz's positional geometry—regarded not individually but in relation to one another. With this analogy, Leibniz bridges the gap between geometry and monadology, between mathematics and ontology. It's a good thing he spared the sensitive aesthete Rémond these complex abstractions; even without moving into mathematics, Leibniz's metaphysics already had him way out of his depth.

A CALCULATOR TO RIVAL GOD

And there's yet another of his many interests that the polymath can't get off his mind this Sunday. Even after the moderately successful demonstration of his calculator in London in early 1673, Leibniz has continued to work on a mechanical apparatus that can reliably carry out all four basic operations of arithmetic. The engineering

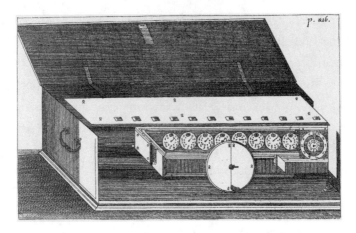

*Leibniz's calculating box. From Johann Christoph
Gottsched, ed.,* Herrn Gottfried Wilhelms Freyherrn
von Leibnitz Theodicee . . . *(Hanover, 1744), insert 3.*

challenges are enormous. The main issue is with the stepped drum,
a gear with teeth of variable length for inputting digits that Leibniz
invented as a young man in Paris, probably inspired by the local art
of clockmaking. The movements of the stepped drum turned by
the hand crank have to be linked to the counting mechanism with
millimeter precision. First in Helmstedt, and now in Zeitz in Saxony, Leibniz has commissioned ongoing work on several models of
his "living abacus." What keeps causing problems is the so-called
decadal carryover, the ability of the counting mechanism to change
over, for example, from 9,999 + 6 to 10,005. Rudolf Christian Wagner in Helmstedt had suggested that in such cases one could simply
reach into the machine and turn the gear by hand. Leibniz had
firmly rejected this solution. He wanted a machine whose operation
was fully automatic, performing calculations without additional
manual intervention.

In Zeitz for quite some time now, Hofdiakon Gottfried Teuber
has been overseeing the work on two models of the calculator, the
older and the younger machine, but he keeps reporting problems
that the craftsman assigned to the task (the *automatopoeus*) can't

solve. Teuber's last letter, from July 30, was once again disappointing in this regard.[31] And yet a *machina arithmetica* of the kind Leibniz envisions could be so practical. Intelligent people wouldn't have to spend their time on tasks that could be performed mechanically. To find a concrete example for how his calculator might be used, all Leibniz has to do is walk out his front door in Vienna. Just around the corner from his lodgings on the Lugeck is a meat market where on certain days poultry, beef, or wild game are sold. A mechanical apparatus for quickly and accurately adding up the meat prices would be a real help.

And so Leibniz doesn't give up hope. The construction of a mechanical calculator has symbolic meaning for him as well. He sees an analogy between man's ability to implant a part of his reasoning ability, namely arithmetic, into inanimate matter—as if the numbers were adding, subtracting, multiplying, and dividing themselves—and God's (far superior) ability to imbue organic machines—that is, living creatures—with intelligence. Human thought and action can be mechanically assisted in any number of ways. Leibniz doesn't try his hand just at building decimal or dyadic calculators; there are also instruments for drawing geometric figures, slide rules, and other analog calculators, even a *machina combinatoria*. He considers it possible that one day humans could build ships that can pilot themselves into a harbor, or automata that can walk clear across town and make turns at certain intersections. He described this last idea in one of his pieces written in Vienna, which is also included in the anthology for Prince Eugene. Leibniz was a visionary of autonomous mechanized locomotion.[32]

Despite such great expectations for the capabilities of technology and mechanical production, he knows there's a limit. We can see God in his creations, he says, but it is difficult for us to understand the perceptive faculties of even so small a creature as the fly.[33] The most complicated machine could never hold a candle to these pesky little nuisances. As simple and straightforward as their small, unconscious perceptions might be, a machine will never be able to imitate them. Here Leibniz stresses above all the indivisibility of mind and body, which poses an insurmountable obstacle to mechanical replication. Feeling, perception, and consciousness cannot be artificially produced—of this he is convinced. Leibniz the machine builder would place a clear limit on present-day expectations for robotics and artificial intelligence.

SOULS WITHOUT WINDOWS?

What a strange turn things can take. On this very weekend, August 25 and 26, with departure in the air and his time devoted to hectic travel preparations, a thought slowly forms in Leibniz's mind: maybe it doesn't make any sense to hurry back to Hanover after all. Where the Welf history project is concerned, he would still have to face his employer empty-handed—not ideal if one is angling to become court historiographer in London. Maybe it's better to wait until George Louis is already in London, then Leibniz can follow him to England later on. On the other hand, to tarry in Vienna any longer wouldn't make a good impression either. He should at least set out for Hanover soon, but he could certainly take his time getting there—he could, for instance, stop for a few days in Zeitz on the way and see how the work on his calculator is coming along. Not a bad idea. Leibniz seems inwardly torn and indecisive, more than ever before. He's heard that the renowned Paracelsus once lived in the building on the Lugeck where he has lodged for so long now, and Wallenstein is said to have conducted astronomical observations from the uppermost part of the tower. The building's name, Leibniz recalls, comes from the city councilman and merchant Georg Federl, who purchased the building in 1590 and added to it.

From his window in the tower of the Federlhof, Leibniz has a sweeping view over the roofs of the buildings clustered tightly together in the narrow streets of the old city. How many times,

looking out, might he have felt lonely or sad? Asked himself where
he actually belongs? Is it in Vienna, where the emperor's court seems
at times close, at others impossibly far away? Or in Hanover, where
he doesn't feel any closer to the court? On this Sunday, the mood
is one of parting. Leibniz's lodgings, like the apartment on the
Lugeck, are places of transition. The manuscripts on his desk reflect
this in-betweenness, this transitory aspect. As ever, Leibniz has
never truly arrived, neither here nor anywhere else. Change seems
to be the only constant in his life. What has long been an expression
of dynamism and mobility is now slowly turning into something
else. Connected to the entire world and yet somehow lonesome and
alone—in Vienna this state of inner division has grown still more
acute. And it shows in the philosophical manuscripts of the last
few months.

The principles and motifs found in these manuscripts are nothing
new: simple substances or monads as mirrors of the universe, the
active inner force, confused and clear perception as a representation
of the many in the one, space as the order of coexisting things, time
as the order of nonconcurrent being, the sufficient reason behind
everything (nothing happens without cause), nature as a continuous,
cohesive whole, the necessity of evil—and the everywhere present,
preestablished harmony within and between all things. Everything,
Leibniz writes in the *Principles*, is thus arranged in the greatest possible order and harmony, once and for all: "The present is pregnant
with that which is to come, the future can be read in the past, the
distant is expressed in the proximate. One could discern the beauty
of the universe in every soul if one could unfold all its layers, which
reveal themselves to the senses only with time."[34]

Nor is it new for Leibniz to think of the world in terms of both
radical individuation and a radical interconnectedness of all individuals. How can something be perfectly self-contained and at
the same time interwoven with the whole world? There has always
been a tension between these two elements in Leibniz's philosophy.
In the writings of the last six months, however, this tension has

intensified, indeed the inner coherence of his philosophy is threatening to split apart. The tug-of-war is visible in two central metaphors of the *Monadology*: the mill and the windowless monad. The
first refers to the physical constitution of the human mind. Let us
assume there is a machine that can feel and think like a human, and
let us imagine it being of such great proportions that a person could
walk inside it as they would a mill. (Maybe Leibniz is thinking
unconsciously of his Harz windmills here.) Inside we would only
find parts that move and interact with one another (in present-day
terminology, perhaps, neurons and synapses); we wouldn't be able
to see the mind itself. The other metaphor reads: "Monads have no
windows through which something might pass in or out."[35]

This statement can be found in similar form in earlier writings,
but here its impact is more resolute and apodictic than ever. What
exactly Leibniz understands this statement to mean is still today a
matter of debate. It points in at least two different directions. First,
at the relationship between body and soul: Leibniz assumes that
there is no direct communication between the two; rather, independently of one another, they find themselves in harmony, like two
clocks that are in perfect sync. In a further sense, his statement concerns the relationship of the individual to the outside world. Here,
likewise, he rejects the possibility of a direct connection, positing
instead that stimuli from without are received indirectly and elicit
an internal response in the form of an expression or representation.
Leibniz once again shows himself to be a philosopher of the indirect
and mediated, of the fragmentarily communicated and the unconscious. With this, he opens the door to a differentiated understanding of human cognition. A connection can be made between the
windowless monad theory and modern approaches to the study of
the brain in which the brain is understood not as something that
reacts to outside stimuli but rather as an organ that, on the basis
of experience stored in the memory, proactively readies itself for
predictable stimuli and only occasionally or partially compares its
actions with current inputs.[36]

As famous as Leibniz's dictum that monads have no windows is, there has been scarcely any research into where he could have gotten the inspiration for such a trenchant notion. No doubt he must have thought of Descartes, who asked himself if all the people out on the street that he saw from his window were mere automata dressed in hats and clothes.[37] Maybe he also thought of a humorous passage in Bayle concerning a certain inquisitive woman on Rue Saint-Honoré who could never peer out her window without her look causing a carriage to drive past.[38] To be sure, windows, or the lack thereof, should be understood here only as figurative stand-ins for the fundamental questions concerning the connection and communication between the self and the world, between the soul and body. But is it possible that Leibniz also has concrete examples in mind? Like the large, colorful stained-glass windows depicting biblical scenes at the nearby St. Stephan's Cathedral? Or maybe he's thinking of the windows in the narrow alleys around the Lugeck, with heavy bars on them to thwart thieves? Or perhaps even the tiny windows in the doors of Vienna's apothecaries, through which medicine was dispensed during the plague months?

Whatever the case, both metaphors have an air of melancholy about them, a sense of lack. One can step inside the mill, but the mind thus enlarged—this one still cannot see. Nothing but boards and beams. Whoever might have come here to observe how thinking and consciousness work is in for a disappointment. And it's even worse with the monads: nothing and no one can get in or out. The self seems inaccessible, trapped within itself for all eternity. Both ideas, then, have a hopeless aspect to them. Everything is connected to everything else, sure, that's all well and good—but *what* is connected to *whom* and *how* seems to remain a question. Monads reflect one another and the world, but they themselves remain strangely invisible, in the same way that, when looking in the mirror, we see what the mirror reflects but not the mirror itself. The idea of a window, one might say, implies the existence of something on the other side, something beyond (even if it's just empty space); the idea of a mirror, by contrast, implies endless depths but no beyond.

Is the world a labyrinthine mirror cabinet, with nothing but darkness and isolation beyond the radiant reflections and flashes of light? Can our souls admit no ray of sunshine, no warm smile? Does it all just reflect off the surface, rebuffed? This notion reveals, indirectly, some of the inner conflict Leibniz has felt in Vienna: he is in contact with many people but still seems somehow alone and forlorn, as if separated from his fellow man by a closed window. The fact that it has been a rainy summer and sunlight has seldom peered through his own window at the Federlhof might well have made this impression even stronger.[39]

However, both motifs, the windowless monad and the mill, are in their own ways hopeful. The unseen mind in the mill remains free; it can neither be locked away inside a machine nor mechanically reproduced. And for something to be windowless implies that it carries the entire world, its abundance, its wealth, *within* itself. Monads have no windows because they themselves are windows. If the only things that existed were God and myself, the entire universe in all its diversity and beauty would still be present within me. Leibniz's efforts ultimately fasten on to this positive understanding. Nothing is more important to him than his own freedom and independence. He wants to serve everyone (particularly those with political power) but be subservient to no one. Ultimately he is striving to not be tied down to any one place, be it Vienna, Hanover, or London. In a letter he wrote yesterday, which presumably is still sitting on his desk at the Federlhof, he offers to serve Charles VI as a kind of ambassador-at-large, shuttling between Vienna and London. Should "my traveling back and forth," Leibniz writes, help establish an enlightening rapport between the two monarchs, he would "consider [himself] happy." Leibniz sees himself as a medium of exchange; he'd like to open a window so that those charged with making decisions that change the world can encounter one another.

TOMORROW NEVER KNOWS

In this sense, Leibniz has remained true to himself throughout his life. Even at an advanced age, he wants to be what he has always wanted to be: a rolling stone moving between the courts of Europe. He is neither a successful power broker, working his influence on high politics behind the scenes, nor a failed, overambitious civil servant. Rather, he can be included among a group of businesspeople, scholars, and artists who over the course of their travels provide the political elite with information. We might call them double agents, not only because they fill this role alongside their actual profession, but also because they do not feel themselves bound to just one side. As informal informants, they facilitate the official business of politics by prompting preliminary discussions of important questions between rival powers, gaining a sense of prevailing opinions, and sounding out political moods, thus preparing the way for those who do bear political responsibility to act.[41] At a time when all communication must be carried out with substantial delays across great distances, this form of political advance scouting is of enormous significance. Leibniz is aware of this fact and tries to utilize the room to maneuver that this role opens up for him as a virtual space of scientific and philosophical freedom. According to this logic, offering his services to several potentates at once is a way to increase his breathing room even further.

But can this possibly end well? In the days following August 26, the news will arrive from Hanover that the soon-to-be-crowned new king will be leaving for London any day now, while Prince-Elector George Augustus (now Prince of Wales) and his wife Caroline Wilhelmine intend to wait a few more weeks before following them.[42] On August 29, after Leibniz learns that Prince Eugene has in fact left Vienna for Baden, he will instruct his servant Hodann not to forward any more letters to him in Vienna, since he "will be setting out as soon as possible."[43] He's in no hurry, however, and makes several stops along the way, including in Zeitz to see Teuber and check on the work on the calculator. Finally arriving in Hanover on September 14, he will neither travel to London with the Princess of Wales's retinue, nor will he ever again return to Vienna. A hoped-for move to Paris the following year falls through, as does a trip to Vienna planned in the summer of 1716 for the following spring. If Leibniz's death hadn't intervened on November 14, 1716, he would in all likelihood have continued to live the roving lifestyle that suited him best. He has spent his whole life on the go. On this Sunday in late August, he is secure in the knowledge that he can face any storm that might blow his way with calm and composure. Let come what may, just so long as he's in motion and never standing still.

CHAPTER 7

HANOVER, JULY 2, 1716

RUNNING HEADLONG INTO THE FUTURE: SPIRAL-SHAPED PROGRESS AND POST-HUMAN INTELLIGENCE

THINKING BEYOND ONE'S SELF

Two great military commanders have big plans for today. In Vienna, Prince Eugene, with the blessing of the emperor and the people of the city, sallies forth, heading south to drive the Ottoman army from the last remnants of Hungarian territory it occupies and from the whole of the Balkans. On the same day, in Schwerin, Czar Peter I makes ready his departure for Denmark, where he plans to gather his strength to strike the decisive blow against Sweden. The two statesmen are bound in opposite directions. Yet they share not only a pronounced thirst for power and conquest but also a weakness for science and culture—traits that make them ideal candidates for Leibniz to petition with his ongoing plans for world optimization. Prince Eugene has not lost touch with the scholarly polymath despite the latter's return to Hanover; the idea of founding an academy of sciences in Vienna is still on the table, ongoing financial difficulties notwithstanding. And the czar, too, is happy to listen to advice from his German Geheimer Justizrat—even if the title Peter bestowed upon him is more symbolic than anything else—and did so in fact just a few days earlier, at the latest in-person meeting between the two of them.

What is Leibniz up to on this July 2? Unsurprisingly, he is sitting in his study at home on Schmiedestraße, poring over manuscripts and books, copying out excerpts, making notes, and writing letters. He knows that it can't go on this way forever. He's not getting any younger, after all. Yesterday he celebrated his seventieth birthday. Day in, day out, lots of things repeat themselves. But the same day, or even one similar to it, won't ever happen twice. Or will it? What if time progressed not linearly but in a circle; what if everything didn't pass away irrevocably but would one day return? On a previous occasion, a year earlier, more in jest than in earnest, Leibniz pictured what it would be like if time revolved in a circle, and he himself turned up multiple times just as he is now: as a scholar, living in Hanover on the Leine, busy with the history of the House of Welf and writing letters to his correspondents all over the world. July 2 is another day that finds him with such things going through his head, a day during which he thinks about what time is, how life developed on Earth, and how it will continue to develop in the future. These are questions he has spent his entire life pondering. Now, however, as he nears the end of his own days—he won't live to see the end of this year—they loom particularly large.

BURSTING WITH LIFE

How long ago his time in Vienna was! In Hanover, the past ruthlessly catches up to Leibniz, specifically the past of the House of Welf, which of course is supposed to be the subject of a monumental work of history—a work that still isn't finished. Any ambitions he might have had to make his stay on the Leine a

brief one have, since the fall of 1714, been systematically snuffed
out by his superiors at the court in England and in the rump
government left behind in the Electorate of Hanover. Premierminister Bernstorff has made it abundantly clear to him that he
must first deliver the history of the Welfs before he can entertain any serious hopes of being named to the historian's post in
London. By this point, his employers don't have much confidence
left. Johann Georg Eckhart, Leibniz's secretary and collabora-
tor in the work of sifting through historical sources, is secretly
being considered as a possible alternative, should Leibniz fail.
Despite clear instructions from the British-Hanoverian government, however, Leibniz doesn't consider himself at all obliged to
neglect his countless other areas of activity. And so he continues
the work on the calculator, publishes reviews and essays, enters
another round in the priority and plagiarism dispute with Newton and his followers, acquires new books, and takes short trips
to Brunswick and Wolfenbüttel.

Only at home is there less variety in his life than before. The
almost-daily coach rides out to Herrenhausen are a thing of the
past. After Sophie's death and the departure of Caroline, now
Princess of Wales, Leibniz has even fewer people to converse with.
But he's by no means bored. Thanks to his epistolary contacts, he
can bring the whole world into his little study. If only he weren't
having these pains. His chronic joint pain and the inflammation
in his legs that makes walking a torment have gotten worse. But
Leibniz doesn't let it get him down. Not being able to leave his
study for long stretches of time isn't the worst thing. To him, it
still feels like he doesn't get to spend enough time in there. "It's
a blessing in disguise," he wrote to Nicolas-François Rémond a
few months ago.[1] One needn't be a philosopher to come up with
a line like that. But it's certainly practical to have handy a philosophy that not only is derived from daily life but can also be
applied to it.

When Leibniz is reading and writing, he forgets about the pain.

He certainly hasn't gotten frail; the pace of his writing hasn't slowed. From time to time, he laments that he won't live to see most of what he plans and envisions. But this doesn't bother him much, and for now at least, talking and writing about these things keeps him alive. For a long time it was thought that Leibniz's productiveness fell off rapidly in the last year of his life, a casualty of his failing strength and increasing debility, but in fact the opposite is the case. The amount of correspondence alone speaks—and fills—volumes. Where older assessments judged the time between 1699 and 1704 to be the high point, in terms of quantity, of Leibniz's correspondence, more recent research paints a different picture. Compare the 872 confirmed letters sent to or from Leibniz in the year 1699 to the estimated 850 to 900 letters sent in 1716. So there can be no talk of a slackening off in the correspondence.[2] On this Thursday as well, Leibniz devotes plenty of time to writing letters, some of them of substantial length. And as ever, their routes to their recipients will take them across borders and in far-flung directions.

The letters written on this day, so far as the record shows, are headed to Theobald Schöttel in Vienna (with an additional letter enclosed, presumably for Schöttel's son Joseph); to the Jesuit priest Ferdinand Orban in Düsseldorf; to Louis Bourguet in Switzerland (more on him later); and finally to the Duke of Modena, Rinaldo d'Este, and to Ludovico Antonio Muratori in Italy.[3] Leibniz conveys various messages, exchanges bits of news, and generally maintains his contacts. In the letter to the emperor's doorkeeper Schöttel, he also touches on mathematical topics, specifically magic squares and magic cubes (cubi magici), of which Leibniz has made a study, as have Joseph Schöttel and the mathematician Augustinus Thomas a Sancto Josepho.[4] Orban receives a request from Leibniz: would he ask the Jesuit missionaries in South America to gather information about plants and animals in Paraguay? Leibniz hopes furthermore that the Jesuits' dispute with the pope over the correct way to spread Christianity in China—the Chinese Rites Controversy—will soon be set aside. Bourguet secures a promise from Leibniz

that he will write a letter of recommendation for a friend of Bourguet's who would like to travel to London. The letters to the Duke
of Modena and to the historian Muratori deal mainly with the
history of the Welfs and the House of Este. For some time now,
Leibniz has been involved in a painful dispute with Muratori over
withheld manuscripts.

TO THE SPRINGS

And so everything seems much the same as usual. Nevertheless, on
this July 2, one matter stands out from the rest of the day's business:
Leibniz's most recent meeting with the Russian czar. He describes
this exciting bit of news at length to Bourguet and Schöttel. The
week before, he met with Peter I at Bad Pyrmont, where Peter
was taking the waters, as well as at the palace at Herrenhausen;
altogether he spoke with the czar several times prior to the latter's
departure last Sunday for the Baltic coast. "I cannot admire enough
the vivacity and the judgment of this great prince," Leibniz writes
to Bourguet. The czar sends for "skillful people from all over, and
when he speaks with them, they are amazed at being so knowledge-
ably spoken to."[5] And in fact, Czar Peter used his stay at the spa to
recruit qualified laborers from Germany: blacksmiths, metalwork-
ers, master builders, weavers, and carpenters, who if the czar had
his way would leave straight from Bad Pyrmont, along with their
families, and settle in Russia.

On July 2, it's all still vivid in Leibniz's memory: the audiences
with high-ranking members of the czar's court; the tasters whose
job was to sample the mineral water from the Pyrmont spring; the
czar himself, complaining about the lack of wine; the entertainment
program with musicians from the Harz Mountains and traveling
performers who led trained bears in a dance before the audience;
the games of chance at the *Kurhaus*; and outside, in the *allée*, the
seesaws shaped like swans and beasts of prey that the young noble

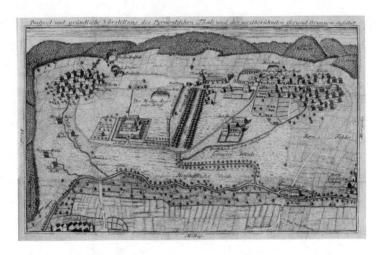

*Map of Bad Pyrmont. The publisher of the map is the
doctor Johann Philipp Seip, whom Leibniz gets to know
in Pyrmont in the summer of 1716 and who will attend to
him on his deathbed in November.* D. Io. Philipp. Seippii,
Fürstl. Waldeckischen Land- und Brunnen-Medici
Neue Beschreibung der Pyrmontischen Gesund-
Brun[n]en *(Hanover, 1717).*

ladies used to pass the time.[6] For Leibniz, it was a welcome change
of pace, less because of the spa town's sundry pleasures than because
of the opportunities—not exactly plentiful of late—to enjoy direct
access to the political elite. His own lord and protector George I
hasn't shown his face in his ancestral homeland since his move to
London. So it's all the more exciting that the czar is here to visit
this summer, allowing Leibniz the chance to present his various
schemes and plans to an influential potentate.

As at his first two meetings with Peter I in Torgau (1711) and
Karlsbad (1712), Leibniz again tries to win the czar's support for
numerous projects. Among them is a research expedition to Siberia
and the Pacific coast of Asia. Leibniz expects that such an expedition might, among other things, take measurements of the Earth's
magnetic field, which in turn, he hopes, might make it possible
to determine the lines of longitude, so important for navigation

at sea. His enthusiasm is undampened when he writes to Bourguet on July 2 that the czar "keeps himself informed about all
the mechanical arts, however, he is most curious about everything
having to do with navigation; as a result he also loves astronomy
and geography. I hope that with his help we shall learn whether
Asia is connected to America."[7] Leibniz's interest in this last subject has a number of different aspects to it: if there is a land link
between northeastern Siberia and Alaska, it would explain how
Noah's descendants were able to reach North and South America.
An important question in the history of human migrations would
thus be cleared up. If no such land bridge exists, one could explore
the possibility of finding a northern sea route through the Arctic
Ocean, through which ships from northern Europe could reach
the Pacific and the lucrative markets of China and southeastern
Asia. Leibniz has never given up on his old idea for an alternative silver route across the Baltic region. His synergetic thinking
begins to pick up steam. In Pyrmont, he promises to send the czar
a model of his calculator soon, which the czar could then give as
a gift to the King of Persia or the Emperor of China. As before,
the most disparate ideas and plans seem to complement each other
almost automatically—enriched by what for Leibniz is the crucial
component for their realization, a prince's power. The meeting at
the Pyrmont springs leaves him feeling as if he'd bathed in the
fountain of youth.

FOR A NEW EUROPE

Over the last few years, Leibniz has come to invest his hopes for a
new Europe in Russia. His interest in the massive empire, seen at
the time as outside Europe, began quite early on. In Peter I's forced
opening of Russia to the West, he sees an immense opportunity for
the progress of human civilization. Whereas at first Leibniz thought
of Russia as functioning as a bridge between Europe and China

(by providing secure trading routes, for instance), it has increasingly moved to the center of Leibniz's global strategy. He repeatedly describes Russia as a *tabula rasa*, meaning that Russia's culture and civilization are malleable; positive achievements from Europe could be brought to Russia without it also inheriting Europe's mistakes and maladies.

Convinced of the czar's commitment to reform, this summer Leibniz once again presents him with a slew of position papers, project outlines, and sketches of plans to reshape every area of society: administration, jurisprudence, the economy, the military, science and research, skilled trades, and education. Leibniz thinks things through systematically, touching on every aspect and every last detail. For instance, in the realm of education, his proposal considers everything from universities and academies of science down to "children's schools" and what they should hope to accomplish. They should, "in addition to being virtue and language schools, also be art schools in which children learn the basics of the arts and sciences. Such schools should also be art schools, moreover, so that children learn a catechism in them, such as an excerpt from the holy scripture, then beyond that something of logic or deductive reasoning, music, arithmetic, drawing, in certain cases also engraving, woodworking, surveying, and aspects of housekeeping, alongside an introduction to the use of weapons and the art of riding, everything in accordance with each child's nature and inclination."[8]

His claims that Russia is a *tabula rasa* notwithstanding, Leibniz by no means regards the country as an empty screen on which to project utopian European fantasies. He is well informed about Russia and its people and is always working to broaden his knowledge—for instance, about Russian geography.[9] He searches for a way to travel from Europe to China that makes use of Russian and Siberian rivers, which in winter could be navigated by sleigh. On top of this, he incessantly studies travelogues and tries to retrace the routes described in them on maps, primarily the map of Asia made by the Dutch geographer Nicolaas Witsen, which was the authoritative map in Leibniz's time.[10] Success and failure go hand in hand in Leibniz's efforts to acquire information. Even though as Geheimer Justizrat he is officially part of the Russian court, most of the time he has only limited access to information from the inner chambers of the czar's power apparatus. It's no different at the springs of Pyrmont, either. Despite the audiences the czar grants him, Peter isn't exactly forthcoming when it comes to letting Leibniz in on his own plans.

Leibniz's vision of Russia as the future of Europe relies on two things: one, on the power of the Russian reformers gathered around the Europhilic czar, and two, on the willingness of Europeans to migrate there, not just skilled laborers like those recruited in Pyrmont, but also, and above all, capable scientists. The Swiss scholar Johann Jacob Scheuchzer, for one, who is currently in search of a position, could be persuaded to take up the newly vacant position of personal physician to the czar. In this role, he could organize naturalist research expeditions to Siberia and help to establish scientific circles in the empire.

But here too Leibniz has to put up with a setback. After tough negotiations with the Russian administration on Scheuchzer's behalf, he finally secures him an appointment in St. Petersburg, only to have Scheuchzer decide he'd rather stay in Zurich. He'd like to study the Alps, he says, which at this time are as much a "virgin territory" as the tundra and steppe of eastern Russia. The warnings of his colleagues in France and England that the customs of the distant land were "raw," "brutal," and "coarse" might certainly have played a role in Scheuchzer's decision. In his letters to Bourguet especially, Leibniz vents his massive disappointment at Scheuchzer's refusal.[11] Leibniz's persistence will eventually pay off, but only posthumously. His plans for an academy plant a seed in Czar Peter's brain, and finally in 1725 (after the czar's death as well), a scientific academy is in fact founded in Russia, with crucial support from European scholars.

At least in terms of presentation and word choice, the Russia plans fit firmly within Leibniz's worldview. In a paper on the restructuring of Russian government ministries written in connection with the meeting in Pyrmont, he counts "His Czarist Majesty" as one of the exemplary sovereign monarchs. As "gods of this world," these monarchs have ordered their governments in accordance with God's own model, who through his "invisible hand [rules] everything in wise and orderly fashion." Demonstrations of this achievement "Europe watches with astonished eyes."[12] Europe must invest in Russia, Leibniz says, in order to make progress itself. The czar's realm seems to him a necessary stage in the divine plan for civilizational advancement. Potential stumbling blocks and obstacles along the way he chooses to see in a positive light. What might seem to many other Europeans as a step backward, owing to their mistaken belief that the giant empire to the east is still on a lower level of societal development, is in reality—of this Leibniz continues to be convinced on this July 2—a great leap forward.

A GOOD MOOD, SPOILED

Euphoria aside, Leibniz has learned to think small, one step at a time. An alliance between Russia and Austria could be one of these steps; bringing European scholars to the czar's court could be another. He writes to Schöttel on July 2 of the meeting in Pyrmont in order to signal to the Viennese court that he still has a good rapport with the Russian ruler, and to Bourguet in the hope that he can help him find a replacement to take up the appointment in St. Petersburg now that Scheuchzer has fallen through. But his lingering high spirits from last week's discussions at the spa are darkened this Thursday by an ongoing controversy. Today he is above all one thing: annoyed. His dispute with the Anglican theologian Samuel Clarke has been dragging on for far too long. It all began last November, when Princess Caroline, still working to represent his interests in London, asked him to state his position on certain philosophical and theological questions. Leibniz, who feared his patroness could end up falling ever more under the sway of Newton and his followers, seized the opportunity to deliver a sweeping blow. His intention was to counter his opponents in the priority and plagiarism dispute over infinitesimal calculus—which was raging uncontrollably—by striking back on the fields of philosophy, physics, and metaphysics. Clarke took up the gauntlet on Newton's behalf and would remain in close contact with the English polymath.

The dispute, which pins one conception of the world against another and is carried out in letters that Caroline has circulated between Hanover and London, has developed into a political spectacle, with half the English court following along, either by reading the two adversaries' letters or hearing them read aloud.[13] The debate centers on issues concerning the metaphysical trinity of God, world, and man. The absoluteness or relativity of time and space is only one of the many points of contention. Each side claims that the other's notion of the world draws a distorted image of divine order. Leibniz's God, according to Newton and Clarke, is not free and has no power to intervene in his creation. Newton's God, per Leibniz, must constantly intervene to correct the workings of the world. Each side accuses the other of having a deformed conception of God. Most of the time they're talking past each other. The whole exchange is like a confrontation between two armies who miss each other on the battlefield and both declare victory afterward.

On July 2, a sullen Leibniz once again summarizes for Bourguet Clarke's objections and his own responses. He has grown tired of the debate, which by this point is no more than a source of highbrow amusement at court. But he can't just drop out, not with the Princess of Wales involved. Without her favor, Leibniz would have to give up all hopes of gaining the position of historian at the English court, let alone that of special envoy between London and Vienna. Still, he's hit on a novel idea, which seems to have come to him a few days earlier, in Pyrmont. There he had the opportunity to make the acquaintance of Robert Erskine (Areskinus), one of the czar's personal physicians—and the timing couldn't have been better. Leibniz comes up with a plan to send Erskine a trumped-up letter containing a Latin account of the controversy with Clarke that could be published in *Acta eruditorum*. In that account, naturally, Leibniz comes off as the clear victor. The plan is not without ulterior motive, for the Scot Erskine is a member of the Royal Society, and has been since 1703, the same year, as it happens, that Newton became president of the society.[14] Yes, that should do nicely. Even at an advanced age, Leibniz clearly hasn't forgotten the tricks and ruses that scholars employ in their intrigues.

POLYMATHS, KINDRED SPIRITS

Amid their quarreling, the two men seem at times to lose sight of the substance of their disagreement. When Leibniz on this July day writes to Bourguet saying that space and time are not absolutes because they cannot be disassociated from things, he is also talking about the age of the world. If the beginning of time coincides with the creation of the world, then it prompts the question of how long ago Earth and the cosmos were created.[15] Leibniz and Bourguet have already had intense discussions on the subject—earnest, probing philosophical discussions, not pithy and polemic exchanges as in the debate with Clarke. Why is he so willing to open up to and confide in Bourguet in particular? In the last few years, the Swiss scholar has become a very close correspondent of Leibniz's.

To be sure, their interaction got off to a rocky start in 1707, when Bourguet came close to linking Father Bouvet's analogy between Chinese hexagrams and binary numbers with heretical Spinozism and in so doing unwittingly criticized Leibniz. On that occasion Bourguet showed himself to be quite capable of touching on a philosophical sore spot. Since then, though, Leibniz has come to know him as a well-rounded scholar whose interests include not only ancient languages and theology, but also philosophy and mathematics, and who is also conversant in a whole range of other disciplines, such as physics, geology, biology, and archaeology. In a three-way discussion with Jacob Hermann concerning Leibniz's dynamics and infinitesimal calculus, Bourguet proved very receptive, and Leibniz has increasingly come to regard him as a fellow polymath with whom he can discuss both scientific and metaphysical matters.[16]

During his most recent stay in Vienna, Leibniz intensified their correspondence even further. At the time Bourguet was in Venice looking after his father's business interests; the elder Bourguet, a wealthy merchant, was a Huguenot who had fled France for Zurich. But now his son, at thirty-eight, has given up his involvement in business and returned to Switzerland, where he plans to pursue a career as a scholar focusing on many of the same fields as Leibniz. Since early 1716, Bourguet has been writing his letters to Leibniz from Morges, a small town on Lake Geneva. The newly minted Swiss citizen regularly provides Leibniz with news from the Italian scientific scene and attempts on his behalf to procure mulberry seeds needed for silkworm cultivation.[17]

In addition to practical matters, the often lengthy letters exchanged by the two men deal with everything that is near and dear to Leibniz's polymathic heart. It's a learned long-distance conversation between intellectual equals, two mental decathletes in dialogue. On July 2 Leibniz is replying to Bourguet's letter of May 15. He takes up various topics that have long dominated their correspondence, but this time he brings them to a conclusion. Several discussions, some of which have stretched out over several years, converge in just a few sentences. It is, so far as the surviving records show, the last letter that Leibniz will write to Bourguet. Bourguet will reply, of course. But there will be no more mention of the topics that have occupied the two of them for so long; they reach their culmination in Leibniz's letter of July 2.

HOW OLD IS THE EARTH?

Where last year's summer was too hot, this year's is too cold. The winter was long, and the cool weather lasted well into May. In Hanover, the wind is still blowing from the east. Most of central Europe is experiencing a very late bloom, and even now, in mid-summer, it rains too often.[18] For Leibniz, who today is still dwelling a little on his impressions from Pyrmont and is also thinking about how he can get that bothersome Clarke off his back, the study of the weather is—no surprise here—yet another area of scientific interest. Dating back to his first years in Hanover, Leibniz has spoken again and again of how impactful a Europe-wide network of weather stations would be. From Scheuchzer, he receives climate data at regular intervals, keeping track of atmospheric pressure, fluctuations in temperature, and precipitation in the Swiss Alps. Long-term changes in climate, as Leibniz is well aware, can be observed only if one is in a position to collect such data over a long period of time and over a large area, then to compare them. Changes in nature and the environment interest him above all from a geological standpoint. They also relate to his questions concerning the age of the earth and how life on it has developed. These questions linger in the background of the time and space debate with Clarke, but in the correspondence with Bourguet, they are addressed explicitly and extensively.

Leibniz considers the question of why God didn't create the world earlier absurd, because time as an independent entity divorced from things is inconceiveable.[19] All well and good. But how old is our world, and when did it come into being? In Leibniz's time, there were essentially two answers to this question: one, the world is eternal and undying, a hypothesis that can be traced back to Heraclitus of Ephesus and Aristotle, and two, it is finite and young. Bourguet, in line with the prevailing opinion in the church and among scientists, works on the assumption that the Earth is just a few thousand years old. Leibniz, however, offers a third answer: the world is finite but old. Fossilized remains and other traces left behind by animals and plants suggest, according to Leibniz, a much longer span of geological time than the common interpretation of biblical chronology allows for. He keeps this thought to himself at first, knowing well that scholars who express such views openly risk losing their position and being accused of atheism.

But in his letters to Bourguet, Leibniz sets caution aside. He emphatically objects to Bourguet's conviction that the formation of many fossils can be traced back to an early geological phase marked by destructive natural forces and major cataclysmic events like Noah's flood. Continental shift and the reshaping of the oceans, the formation of mountains, and the fusing of rock did not, Leibniz believes, happen suddenly and abruptly but rather in a process of slow, even development, which required a very long timeframe beyond the brief span of historical time. Where Bourguet sees catastrophic disruptions at work, Leibniz perceives uniform forces acting both in the present and in the geological past. The same principle of continuity that he argues for elsewhere, a principle that ensures seamless connection and fluid transitions in nature, he applies to geological processes as well. Creation itself he interprets as an ongoing process (*creatio continua*), which he expresses in concrete terms as the formation of planets out of a succession of solar eruptions.[20]

In the Leibniz-Bourguet correspondence, the relatively static worldview based on biblical chronology transforms into a dynamic model of geological time that in theory could be extended indefinitely. It's not that Leibniz is the brilliant exception among scientists; rather, he takes his place among a group of other contemporary scholars, like Niels Stensen (Nicolaus Steno), Robert Hooke, and Benoît de Maillet, who no longer feel bound to a timeframe of a few thousand years. Bourguet, too, is influenced by his discussions with Leibniz and will not return to the old schema in the works on geology and paleontology that he will author many years from now.

Together, their voices, which herald the divergence of the biblical and the geological reckoning of time, form an epistemic foundation on which a new paradigm of geological time will establish itself in the late eighteenth century, at which point it will no longer seem impossible to place the age of the earth at tens of thousands, hundreds of thousands, or indeed even millions of years.

THE SALAMANDER
WHO CAME IN FROM THE HEAT

The letter of July 2 has a bit more to say about natural processes that unfold over a long period of time. There's also the matter of the development of organic life over the course of Earth's history. Leibniz sees evidence of the ongoing development almost daily, at least when it comes to plants. He visits his garden by the Aegidian Gate more frequently than he used to. There's no record of whether he goes to the garden this Thursday, but these days he is often observed making his way there, gawked at by the children on account of his large, pitch-black wig and the ancient patched clothes he wears ("rags," Electress Sophie calls them), traveling sometimes on foot, sometimes in his carriage, with its velvet upholstery, its outside painted with a floral design against a coffee-colored background, and its interior always, apparently, full of old junk. There's more in Leibniz's garden than just mulberry trees for silkworm cultivation. Much more, in fact: there's a fertile field of fruit and vegetables. Leibniz has had cauliflower seeds sent over from Holland; there are peaches and several varieties of apples. Up to seven people are employed in the garden, tilling, plowing, and planting or tending to the crops. Despite the rainy summer, the garden plot is in full splendor by early July. Green thumb or no, Leibniz also has an eye for the Earth's creeping things, above all silkworms—hatching from their eggs, eating, spinning, and metamorphosing into moths. This interest in insects doesn't escape the curious children either.[21]

Where did these little creatures come from? How did they come to be? Leibniz rejects the theory, which dates back to Aristotle, that they were spawned from rot or slime. Once again a certain tiny pest is called upon to serve as an example. At one point, it was believed that honey flies (*mouches à miel*) were spawned from the flesh of bulls, Leibniz writes. Now, however, we know that certain large flies or wasps that are very similar to honey flies bore into the leathery skin of cattle and lay their eggs underneath.[22] Life doesn't form anew; rather, it is always already present. This hypothesis, as we can see, unites Leibniz's theories of preformation and of the monad. Whether bodies and souls are to be found in male semen, as Antoni van Leeuwenhoek claims, or in the female egg, as posited by Antonio Vallisneri (also Vallisnieri), is, however, an open question for Leibniz, and one he continues to discuss at length with Bourguet. In August 1715, Leibniz starts up a correspondence with Leeuwenhoek, the Dutch scholar of microscopy, whom he met in person almost forty years ago, for the express purpose of addressing this question. And through Bourguet he establishes contact with the Italian naturalist Vallisneri, who promises exciting new discoveries in this field. Leeuwenhoek reports to Leibniz that he has built models of flies, fleas, and gnats out of copper wires in order to better understand how the sinews and muscles of the tiny insects function. On more than one occasion, Leibniz will proclaim, in reference to such empirical experiments and observations of microscopic nature, that he values what a Leeuwenhoek sees more than what a Descartes thinks.[23]

Bourguet's conjecture that flies and gnats have more eggs so that they can serve as a source of sustenance for individuals of the species without the species as a whole dying out leads to the question of whether species can change over time. The official line sanctioned by church and state strictly denies that species change. And here too Leibniz is willing to break the taboo. He posits that the Earth started out as a glowing hot ball, before it was completely covered in water. Like the naturalist and theologian Niels Stensen, with whom he was able to discuss this and other subjects in person in Hanover between 1677 and 1680, Leibniz assumes that amphibians, and subsequently land animals and birds, developed from fish.[24] In his correspondence with Bourguet, Leibniz hits on the idea that "from salamanders (so to speak) fish could have come to be, then amphibians, and finally land animals and birds."[25]

With this curious idea, Leibniz is alluding to the myth, widespread since antiquity, that salamanders can live in fire (hence the name "fire salamander"), and that indeed this is their natural habitat. Maybe Leibniz isn't being entirely serious when he voices this thought, but he is quite unmistakably playing with the idea of the evolution of species or life-forms and is thereby venturing far outside the boundaries of what can permissibly be said in his time. Nor does he rule out the possibility that individual species may have gone extinct. In the mountain caves of the Harz, he has seen bones that can't be said to belong to any living species of animal. The remains of horns of the legendary unicorn, Leibniz conjectures, could have belonged to fish from the Southern Ocean. He doesn't think much of the conclusion that fossils recently found near Quedlinburg could be the remains of a unicorn skeleton. Nevertheless, in 1716 he commissions Nicolaus Seeländer to make an engraving depicting the so-called Quedlinburger Unicorn to use as an illustration in his planned study of geology, the "Protogaea" (primordial earth).[26] Still, if Leibniz was an early exponent of the idea of the evolution of species, as it will later be propounded by Charles Darwin and others, it's only to a limited extent. For him, the agents driving the evolution of species are not contingency or natural selection, but rather a divine plan that has predetermined the development of life-forms in accordance with changing environmental conditions. Leibniz's evolution is guided not by blind chance but by a divine *telos*.

If life-forms and species evolved from other life-forms and species, how exactly might such a development have taken place? "Death could bring the salamander back," Leibniz writes to Bourguet, "and perhaps in a higher state of refinement than before."[27]

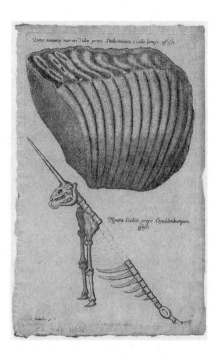

"Quedlinburg Unicorn"
and molar fossil of a woolly
mammoth. Gottfried
Wilhelm Leibniz,
Protogaea . . . *(Göttingen,*
1751), pl. 12.

If birth and death are merely moments of transition in the progressive and regressive development of life, then certain individuals must be able to undergo transformation and be elevated to a level of greater refinement—which is to say, to become a more highly developed species. Following this line of thinking, Bourguet concludes that every individual organism undergoes a transformation from sperm to animal and finally to creature endowed with reason. Leibniz objects that this is by no means necessary. How exactly the reshapings led to the emergence of new life-forms and species, however, remains an unsolved problem for him. But he doesn't linger on such issues. Instead he goes looking for a general concept of substance that includes both inorganic and organic matter. He posits that there must be coherent entities that can organize and stabilize themselves and that are capable of reproducing by taking in energy from the environment—or so Leibniz's thoughts might be reformulated to draw a connection between them and modern approaches in biology and chemistry.

Scientists in these fields now propose that such autopoietic and self-replicating entities must exist even where no fully developed organic substance is yet present. Otherwise the evolutionary processes of variation and selection couldn't take hold and get going. Such hypotheses and attempts at a general theory of evolution are still in their infancy, but should their story ever be told, Leibniz certainly shouldn't be left out of it.[28]

Leibniz considers these reflections so significant that he includes a copy of his May 1714 letter to Bourguet in the anthology he gives to Prince Eugene under the title *Lettre sur les changements du globe de la terre* (*Letter on the Changes of the Earth*). But because Leibniz forgot to list the letter in the table of contents—maybe the hectic departure from Vienna in late August 1714 was to blame—it has to this point been largely overlooked in Leibniz scholarship.

CAN THERE HAVE BEEN A BEGINNING?

All these lively and passionate discussions happened more than two years ago. But the traces they left are impossible to miss in the letter Leibniz addresses to Bourguet on July 2; one need only read between the lines. The key concept in this letter is infinity. It forms the link between natural history and geology, the space and time debate (in which he is enmeshed with the bothersome Englishmen), and one more issue that will also come up for discussion here. Again Bourguet gets the ball rolling by asking Leibniz if there necessarily has to have been a first moment. No, says Leibniz, time, unlike numbers or logical propositions, cannot be broken down into fundamental units or concepts. Time doesn't reflect anything necessary; it reflects only the course of events in the world, determined by many random occurrences. Thus for logical reasons, there doesn't necessarily have to have been a starting point. According to Leibniz, either the world has always been unvaryingly perfect, which would rule out the possibility of a beginning, or the world is increasing in perfection, with that growth taking either a linear or a curvilinear form. In the curvilinear model, the world would exist eternally, because time would progress asymptotically—that is, it would approach a beginning and end and get infinitesimally closer to them without ever reaching the one or the other. Only the linear progress model has a beginning, Leibniz says. He sees no way to decide on one of the three hypotheses based on logical reasoning or a proof.[29] So the world is either perpetually good or it's getting better. Leibniz rules out the possibility that it could be getting worse. The notion, for instance, that there was once a Golden Age of the Germans that is now lost—this Leibniz rejects, along with similar theories of decline.[30]

Bourguet, who eagerly takes up the discussion, decides on the third model: if the universe has a divine goal toward which everything is moving in the form of a rising line, then there must have been a starting point. Leibniz won't commit himself, even as the debate stretches out over several letters. He likely does consider it probable that there was a first instant. It's even possible, he says, that other universes existed before our own and that more could follow after ours. But he doesn't want to fundamentally rule out the possibility of a world without a beginning. No, he writes, he's going to stick with his present, ambivalent position. After sufficient meditation, he has come to the conclusion that the matter cannot be proven one way or another.[31] To decide, after ample reflection, not to decide is typical of Leibniz. He sees in this a kind of economizing principle for intellectual resources. Taking on numerous tasks at the same time leaves him with notoriously little time for drawn-out reasoning or reckoning. His approach is "like that of a tiger, of which it is said, that what it does not catch with its first, second, or third pounce, it lets go."[32]

The letter of July 2 sounds the final chord of the debate. Leibniz continues to maintain the fundamental indeterminability of the beginning problem, though he concedes that the unalterable perfection hypothesis seems least plausible to him. In that scenario, the world would always have been coexistent with God at the highest level of perfection, without temporal bounds. God and the world would therefore exist co-eternally (*coeternelles*).[33] By contrast, he suspects that the world and its inhabitants have a finite past (albeit one that goes a long way back) but possibly an infinite future. As much as the world might increase in perfection, divine perfection, the highest possible form, remains unattainable. The idea of God's having chosen the best possible world refers not to any one temporally bounded period over the course of the history of the universe, but rather to the entire sequence of things and events—the most perfect of all possible sequences. In God, the idea of the work always precedes the work itself, from the beginning on into eternity. Leibniz expressed a similar view in the dispute with Clarke, although not in such a philosophically impartial manner. To Bourguet, meanwhile, he candidly admits that ultimately he cannot prove his hypotheses, either with pure reason or with rigorous logic.

THE RETURN OF YESTERDAY

And so today, once again, Leibniz sits in his study on Schmiedestraße, armed with quill, ink, and paper, writing and thinking the thoughts that he has written and thought so many times before. It's almost like Groundhog Day. But even if Leibniz doesn't manage to come up with proof for the theory that the world is growing steadily toward a higher degree of perfection, he finds a different way of substantiating his views, one that likewise leads us to these days in early July 1716. For parallel to the time-and-infinity debate he's conducting with Bourguet, he is having a similar discussion in a literary context. Sitting on his desk at the moment are some fairly large piles of paper containing another author's writings, which he is looking through and correcting. The text in question is a long manuscript by Johann Wilhelm Petersen, a poem that runs to more than fourteen hundred lines. Leibniz has promised the author, who was once the superintendent of Lüneburg but was stripped of his church office in 1692 on account of his nonconformist millenarian views, that he would edit the voluminous epic before its publication. Now, at long last, he is finally finished and will be handing the edited manuscript, which bears the title *Uranias*, over to a mutual friend of his and Petersen's, Johann Fabricius, when he travels to Wolfenbüttel in a few days.[34]

In his work, written in a poetic and popular style, Petersen wrestles with the historical and religious question of whether the fallen state of the world can be ended and the state before the fall restored, a return to Eden that would include overcoming the punishment of damnation and reversing the angel Lucifer's fall from grace. Viewed from this position, Leibniz's eternity problem takes on a new aspect. Infinity can be seen not only as an asymptotical curve of progress, or as a constant—or at least minimally wavering—invariability, but also, alternatively, as the cyclical return of eternal sameness, somewhat along the lines of the cosmological teachings of Origen, the theologian of antiquity, which Petersen also touches on. If history behaves like other periodic natural cycles, then it cannot rise in linear fashion, as in the model that Bourguet believes to be plausible. Shortly before he gets into the beginning-and-infinity discussion with Bourguet, Leibniz starts to discuss Petersen's cyclical salvational theory with Adolph-Theobald Overbeck, deputy rector of the *Gymnasium* in Wolfenbüttel.[35] Here then are two discussions that begin at only a slight temporal remove, gather momentum at the same time, and finally coincide and run together on July 2.

UNDERSTANDING THE FLIES

The essence of the exchange with Overbeck is distilled into a short fragment later known under the title *Apokatastasis panton* (Universal restoration).[36] Leibniz appears fascinated by the thought that life as a whole could act like an hourglass that keeps running out and being flipped over again and again—a loop of perpetual repetition that goes so far as to apply to individuals leading a succession of identical lives. Here too Leibniz expresses that half-serious notion—mentioned earlier—that he himself could one day return as a scholar living in the city of Hanover on the Leine River, occupied with the history of Brunswick and penning letters to his friends.[37]

Unlike Petersen, Leibniz does not believe in a recurrence of history and attempts in the *Apokastasis* text to refute this claim and win Overbeck's agreement. In the course of his argument, he goes along with Petersen's idea at first, hypothetically. Assuming that everything that is bound to become reality will have happened once at some point, and that, at the same time, everything will have been written about it, then reality would eventually be exhausted, and history would have to start over from the beginning. In order to demonstrate the repeatability of history, down to the actions of individual persons, Leibniz comes up with a combinatorial thought experiment: given a finite number of letters that every year are combined to produce a record of human history, the result would be a likewise finite number of possible books recording histories that, within a suitably large timeframe, would have to repeat at some point. The library that would house such books would be gigantic, but ultimately its dimensions would be finite.[38]

After presenting this notion, Leibniz scraps it, pointing to the unnoticeable changes that cannot be perceived by the senses and cannot be set down in any book. He refers to his theory of minute perceptions, that is, the singular perceptive states unique to all simple substances, in which everything in the universe is reflected. Ultimately the *petites perceptions* ensure that history can continue to advance outside our notice. For the most part, this occurs in spiraling fashion. We sometimes take a step back, only to take a great leap forward. And this theory of progress also applies to human cognitive faculties. Allowing for their development over extremely long periods of time, these faculties could one day grow so advanced that everything collected in the universal library Leibniz imagines could be expressed as a concise formula that would fit on a single sheet of paper.

A condensed world formula from which all future knowledge could be derived: this notion presupposes a cognitive apparatus, a highly developed mind, that has advanced over a very long span of time beyond the present constitution of human reason. With this hypothesis of a post-human *ratio*, which of course in Leibniz's metaphysics still does not converge with the divine, humans are granted logical entry into the teleological narrative of evolution, fitted into the long chain of transformations and metamorphoses: salamanders, amphibians, land animals, and birds; humans endowed with reason and post-human intellect. What exactly this advanced reason will look like must, however, remain hidden to today's powers of understanding—Leibniz dampens any hope of an overly hasty anticipation of what at present cannot be imagined. Still, he does at least give us an example of the massive capabilities of such minds. They would be able to comprehend at a glance not only the definition of a fly, which is infinitely more complicated than that of a circle, but also the substantially longer definition of a particular species of fly, to say nothing of the almost immeasurably complex formulae pertaining to an individual specimen of fly.[39] Or to put it another way: the example of so harmless and tiny a creature as the fly, of all things, demonstrates that any attempt at a complete description of the world under the present conditions of human rational capacity is doomed.

A DAY IN THE LIFE

The limited nature of the human capacity for understanding is also the reason Leibniz still continues to maintain, on July 2, 1716, that it is impossible to present any clear proof that would resolve the question of the beginning and end of time, the universe, and the history of the Earth. Even if everything had already been said before—and by everyone—that still wouldn't be the end of it, not even close. Of all things, combinatorial analysis—Leibniz's great hope for providing a comprehensive description of the world, past, present, and future—runs up against its limits here. The same goes for the dyadic, into which such a description might possibly have been parsed out and expressed. How can this be? Is Leibniz contradicting himself here?

On the one hand, he posits the notion of a comprehensive representation of all phenomena by means of a tightly woven network of characters and symbols, perhaps taking the form of a never-ending cascade of zeroes and ones, together with a formula or an algorithm that would make it possible to calculate the knowledge of tomorrow. On the other hand, there are the materially intangible minutiae of a ludicrously large number of simple and immortal substances that incessantly reflect one another, plus a limitless flood of immeasurable waves of unconscious perceptions that perpetually permeate the entire cosmos. The unpredictable; the tiniest of chance occurrences with the greatest possible effects: the famous flutter of the butterfly's wings that triggers a hurricane, or the buzzing of a fly that, however intangibly, travels to the other end of the universe. Two irreconcilable, mutually exclusive notions seem to be colliding: the total scientific comprehension of the world versus the fundamental inadequacy of any attempt, in whatever form, to comprehend. Is it possible that unity and plurality simply cannot be reconciled? What's true on a large scale is true on a small scale as well: for a long time now, Leibniz has been unable to fit everything on his mind under the all-encompassing roof of a consistent metaphysics. Is his scholarly overextension, as evidenced by the sea of stray slips of paper surrounding him, not representative of the idea that beyond every piece of writing-filled paper, the endless sphere of what cannot be expressed in words only begins? In short: is it Leibniz against Leibniz? In a certain respect, sure, only for him that's not a problem. He has no trouble thinking beyond himself as a human individual, as well as beyond humanity as such. What today is indescribable, tomorrow we might be able to spell out effortlessly, even if the numinous—or in secular terms, the nameless—will always remain just beyond our reach.

Leibniz casts his line out far. But among his many correspondents, Bourguet, for one, isn't biting; he doesn't take up the discussion again. His reply to Leibniz's letter of July 2 is full of the usual scholarly small talk—Bourguet mentions the meeting with the czar in Pyrmont, agrees with the arguments against Clarke, gets worked up once again over the plagiarism accusations made by Newton's followers, and touches on a few other things. But for him the great debate on time and space, life and death, history and the future seems over.[40] Like Leibniz, he knows that for now it's impossible, and won't be possible for a long time yet, to say the last word on these fundamental questions, but life carries on all the same. The show must go on. For Leibniz, it's no different.

For of course he has other interests flying underneath the wide-reaching radar of his metaphysical meditations on the nature of the world; he is also thinking about quite mundane, practical things on this July 2. Thankfully, since the founding of a new state bank in Vienna, he has been regularly receiving his Reichshofrat's salary, even if it remains unclear with each payment whether the money should be understood as a paycheck or a pension. Leibniz would prefer the former, so he wouldn't have to worry about any portion of the money being taken out for taxes—and even more important, Reichshofrat would be more than just an honorary title; it would place him on equal political footing with the other properly remunerated Reichshofräte.

Leibniz's financial interests in the empire's capital seem to be in good hands with the imperial doorkeeper and mathematician Schöttel. Just recently Leibniz sent him an authorization empowering him to secure a higher interest rate at the bank on his behalf. Now, today, he asks Schöttel about the possibility of transferring more money from Hanover to Vienna, "in order to deposit said money in the city bank and earn interest on it. The question is, how best to do this."[41]

Clearly Leibniz is entertaining the idea of going back to Vienna soon, maybe even settling down there for an extended period. A further indicator is that he is keeping an eye out for potential lodgings in the city. Abbot Giuseppi Spedazzi, secretary in the imperial decoding office, another of Leibniz's acquaintances who is looking out for his interests, would even be willing to share his rooms with him. Leibniz is also looking to buy a small piece of land in Hungary so as to be able to procure costly wood, hay, and oats for his horses. But the asking price strikes him as too high.

Leibniz probably won't ever be able to give it up: in his mind, he still aspires to the life of a scholarly nomad—or at least he continues to dream of it; he just can't help it. Old age doesn't guard against foolishness; nor does it stop him from keeping the ideas of his younger days alive. Leibniz's "foible," as he has declared on countless occasions, is science, and the life he leads, we might add, is that of a rolling stone. Thus everything seems to be the same as ever. Come evening, it doesn't get dark until quite late. Most of the houseflies settle down to rest, except in those places where a light still burns—in Leibniz's study, for example. Tomorrow will be yet another new day. One could almost say that if Leibniz hadn't died, he would still today be . . .

*Leibniz to Jacob Hermann, November 2, 1716
(excerpt). This letter is believed to be the last
dated in Leibniz's own hand.*

EPILOGUE

HOW THEN TO BID GOODBYE TO LEIBNIZ? REALLY THERE can be no goodbye, for his thoughts live on for as long as we continue to wrestle with them. In October 1716, his health deteriorates drastically. The painful inflammation in his arm and leg joints leaves him increasingly confined to his bed. By the end of the month, he can hardly walk or write anymore. In early November he pens his final letters.

He stops writing on November 6, as reported by his amanuensis Johann Hermann Vogler.[1] A few days later he must also stop reading. On the evening of November 14, a Saturday, Leibniz dies in his rooms on Schmiedestraße in Hanover, probably of acute cardiovascular failure. According to Vogler, shortly before he dies, surrounded by many books, manuscripts, and letters, he reaches for paper.[2] It takes a while for word to get around that Leibniz is no longer alive. After his death, travelers continue to come to Hanover to seek him out. The same goes for letters addressed to him; the last one, as far as the record shows, comes from Johann Bernoulli, written on December 5, 1716, and sent from Basel.[3] Direct communication with Leibniz ceases, but not the conversation with the countless writings, drafts of letters, and notes that he left behind, all of which survive because, in accordance with a common practice for court officials at the time, his massive trove of manuscripts in Hanover is sealed and preserved upon his death.

So what remains of Leibniz? There are his sensational breakthroughs in the sciences, the foundation of modern calculus being just one example; then there are his many technical inventions, such as the stepped drum for his calculator or the looped chain used in mining; and beyond that his countless practical proposals for the reorganization of education, science, administration, and the economy, to say nothing of his work elaborating on basic theoretical principles and recurring concepts without which the modern scientific world is unimaginable, for example the relativity of space, the law of the conservation of energy, and the binary numeral system that he so eagerly promoted. And in addition to these innumerable individual achievements, there are the ideas—albeit only vague and speculative—that will be developed to their full extent only in later years: a wide-ranging idea of evolution that also encompasses consciousness and intellect, his struggle to find a general concept of substance that goes beyond the divide between organic and inorganic matter, and more.

In his unflagging ambition to link together all these individual bits of knowledge and situate them within a collective frame of reference, Leibniz probably remains an exceptional thinker to this day. His overflowing creativity, constrained though it is by the corset of rigorous rationality, is characterized above all by his having attempted several things at once and all of them related, to wit: (1) to delve into and grasp all contemporary areas of knowledge at once, in both theory and practice; (2) to further develop each in a way that is both *tangibly* innovative, such as by coming up with new inventions, and also *methodically* innovative, such as by regarding a given issue on a higher plane; (3) to carry over insights from one field to another by sussing out analogies and correspondences; (4) to generate in this manner new areas of knowledge—in part by linking together heretofore unconnected fields; (5) to bring them all under a systematic comprehensive framework by tracing them back to general fundamental principles; and finally, (6) not only to formulate gains in knowledge from the aforementioned points abstractly (or sketch them out graphically) but also to give them a concrete material form.

The prime example of the last is Leibniz's calculator. The so-called younger machine, which operated on the basis of the decimal system and was meant to carry out the four basic arithmetical operations automatically by means of a hand crank and a stepped drum, has survived, in restored form, having gone missing after Leibniz's death and at some point being disassembled into its component parts. Today we know that it works; Leibniz, however, fretted over its functionality till the day he died, receiving steady reports from Zeitz, where the machine was being worked on, claiming that there were still numerous difficulties.[4] Even if Leibniz was in many of his endeavors only partly successful, he can in fact be called a "universal genius," to the extent that one can divest this term of the sense of romantic exaltation it took on in the nineteenth century and still partly carries today. Only in a specific sense, however, was he the "last" polymath or universal scholar, as he is often described. He is more precisely characterized as one of the last representatives of a certain type of universal scientist, a type distinguished by the seventeenth-century scientific revolution's optimistic trust in progress, as well as by the hope, dating to the Renaissance, of being able to lay claim to all the world's knowledge in encyclopedic form, to analyze and comprehend it fully.

To describe him as a court scholar, as a typical yet exceptional exemplar of the Baroque adviser, is also thoroughly fitting. But he was never an obsequious courtier. A telling example of this is the history of the Welf dynasty. The duke who commissioned it hoped it would be a song of praise, but in Leibniz's hands, it quickly mutated into a source-critical work of history. The line between autonomy and opportunism was a fine one to walk, but Leibniz kept hoping that by means of his constant oscillation between the power centers of Europe, he could carve out a free space of self-determined, independent study. To know it all, to be capable of doing it all, to set it all in motion—in a certain respect, no one since Leibniz has attempted anything comparable. Neither Kant nor Hegel revolutionized mathematics or experimented with windmills alongside their brilliant philosophical work; nor did Humboldt delve deeply into every area of philosophy alongside his groundbreaking research in the natural sciences.

What else is left? More even than the glowing achievements, perhaps the failures, the resistance, the defeats. Why consider them? Not only because in them we see the differentiation of the sciences that was already advancing in Leibniz's time, as a result of which even an exceptional talent like his was no longer able to maintain a command of all fields; not only because science would be unthinkable without trial and error; but also, more than anything else, because the setbacks he faced bring out Leibniz's characteristic optimism. Never give up, keep going, see the positive even in the negative: for him these aren't just words to live by. He took this attitude and made of it an explanatory principle of global proportions; he turned it into a worldview. God knows not everything has a bright side, for which reason the mockery leveled at him in Voltaire's *Candide* wasn't entirely unjustified. Leibniz's faith in progress, too, is thoroughly problematic. It's true that, unlike many figures of the late eighteenth-century Enlightenment, he doesn't posit a linear form of progress—rather, for him it moves in the form of a spiral—but he too remains largely blind to the fact that nature is not an inexhaustible resource for human global optimization.

Still, what makes Leibniz's optimism attractive to this day is his determination to think in terms of possibilities, of options. The pessimist knows of only one direction that the world can move in, whereas the optimist sees alternatives and opportunities, recognizes potential. Leibniz needs the concept of the possible, not only to enable the selection of the best but also, in equal measure, for the appeal he makes to his fellow citizens to participate in shaping world events. Lots of things may be improbable, but they remain possible, or at least not out of the question. The attempt is always worth it. This, at the very minimum, is what Leibniz offers in his rejection of pessimism.

In this sense, he doesn't merely represent a lost world of controlled, well-ordered optimism, an optimism that at times asks far too much of rationality and temporal confidence. He also occupies a position at the beginning of a new age, maybe even a new epoch, that doesn't yet have a name because it's still ongoing. Leibniz's thinking was usually a "yes, but." Yes, Descartes was right, we can explain most things in the world with the help of mechanics—but a purely mechanical understanding of the world comes up short. Leibniz, of all people, the passionate and gifted inventor of machines, sees the natural as being superior to any product of artifice. Don't stop with the "moderns," the Cartesians and their ilk; go beyond them by also looking back and drawing on the philosophy of Aristotle. Leibniz's endeavor to never come to rest on a single thought paradigm, no matter how convincing it might be, but rather to constantly question and reevaluate it—from a different perspective or at a meta-level—isn't useful just when directed at classical mechanics from Descartes to Newton. Conceived as a general principle, it can also be extended to the theory of relativity, quantum mechanics, or any other paradigm that's still to come, in physics and in other sciences as well. This is the formal message of Leibnizian *meta*physics, which, from a heuristic standpoint, has continued to remain highly relevant in a scientific world that largely understands itself in post-metaphysical terms.

However one might view the relationship between metaphysics and modernity, in Leibniz's case at least it becomes clear that in "Western" modernity, inconsistencies and contradictions are observable from the beginning. The worldwide dissemination of Christianity seems like a promising idea only if the main priority of the project is reduced to spreading a natural religion of reason. The claim to universality and the processes of globalization gradually diverge, even if Leibniz's thought is still far removed from the imperialist, colonialist Eurocentrism of the nineteenth century. It's true that knowledge and the world are still in overall agreement; there are no truths, plural, only infinitely many ways of looking at one and the same truth. But the first tiny cracks and fissures are already visible in this understanding of the world. In considering the late Leibniz especially, we see an increasing tension between radical individuality and the jointly shared worldview that constrains it. That he nevertheless attempts to bring the two into harmony not only demonstrates a deep ambivalence within Euro-American modernity but also shows that the attempt to overcome such rifts is inherent within this modernity.

One more thing that Leibniz makes available to us: Is there really nothing in the world but a mutual mirroring and reflecting, an observation and contemplation of the first and second order—nothing, that is, except that which makes this reflecting and observation possible? And is that what it comes down to—what it all comes down to? Or is it all just a construct? Is the world a digital calculator? Leibniz didn't invent the idea that one should imagine different possibilities, but he did raise it to the level of systematic philosophical category, and in so doing, he also influenced the discussion around virtual worlds. Simulating different worlds, models, and future scenarios is an everyday activity in countless scientific disciplines. The notion that we all just live in a gigantic computer simulation keeps popping up again at again at regular intervals. Whether in science or science fiction, the theory that someone—some sinister mind (*genius malignus*) or representative of a technologically advanced, post-human species—is tricking us with a highly complex yet simulated reality presupposes that there must be an "actual" world. Otherwise it makes no sense to speak of simulations, since it wouldn't be possible to distinguish them from the real. Just as, in Leibniz's dream fable, Theodorus was able to see many projected worlds with many different Sextuses in the Palace of Fates, but only one world with the actual Sextus. Why is there a world and not no world? Leibniz's answer may no longer be convincing, but his question is still compelling. Only if the world were to consist of the most minute units of information, as many believe today, only then would it be possible to reproduce it. So far there is no proof for this; that hasn't changed since Leibniz's time. He had hoped for the possibility of a comprehensive description of the world, though also remained skeptical and pointed to the endless number of unconscious perceptions. For him, the last word on the matter had yet to be spoken; the answer was in the stars—or in tiny, submicroscopic particles.

Either way, Leibniz can be considered one of the scientists who discovered microscopic worlds. He opened philosophers' eyes to the very small, and he enjoyed looking at the world through a microscope even more than through a telescope. Some would later claim that he could never hurt a fly because he considered flies to be so artfully devised, divine machines fitting everything necessary for life into the smallest package, while an elephant requires an exponentially larger amount of space for the same purpose.[5] Still, despite his fascination with the endless reaches of the microcosmos, Leibniz never lost sight of the fact that even the artificially enhanced gaze cannot overcome the limits of sensory perception.

Everything is reflected in everything we do. Thus seven days are enough to understand how Leibniz's life and thought are connected and how, in different phases, each helps shape the other. Seven days are also enough to see that Leibniz is capable of tuning out everything around himself and concentrating only on his abstract thoughts. It doesn't matter what situation he finds himself in. Whether it's Paris in the fall, winter in the Harz Mountains, spring in Berlin, or summer in Vienna, his philosophy thrives in equal measure in its moments of dialogue with others and in its radical withdrawal into the self. And this very duality is expressed in his conception of the world. Everything is connected to everything else. In the consciousness of an individual, every day is linked with every other day by means of distinct memories, but also of confused impressions that have a lasting effect on the unconscious. And there is, to yield the floor to Leibniz once more, no bit of matter, however minuscule, that does not have present within it a world of infinitely many creatures, and no individual substance, however seemingly insignificant, that does not exert an influence on all other substances and that is not influenced by them in turn.

Maybe this is the thought that most connects us, who live in the present, with Leibniz, precisely because it seems so unsurpassably radical—abstract and vague, impossible to really prove, but impossible to refute, either. It's just a part of the modern world somehow, this thought—there's no doing without it. Hidden, it continues to work its influence and ensure that the world remains a bit enchanted. Was Leibniz the first and last magician of the modern age? Presumably he would have had little use for such a label. What counted for him was that the complexity of the world could be resolved and grasped only to the extent that (human) cognitive capacities increased—though we who inhabit this world would probably never completely awaken from our perceptive slumber. Still, he gives a word of encouragement here, too, of course. Not even God, strictly speaking, is able to exactly predetermine the infinite number of contingent truths in every possible world; nor can he round off an equation containing infinite values or series of numbers. He can only take in the totality of them with his unfailing gaze. We can take some comfort in this, at least, comfort that remains available to a secular world. We don't have to understand what infinitely small numbers are; it's enough to know how to calculate with them.

And here at the end, the pen is lifted from the page—and all the possibilities are open.

ACKNOWLEDGMENTS

I THANK TANJA HERMANN AT S. FISCHER VERLAG AND THE whole team there for a wonderful collaboration. My thanks also to the staff of the Leibniz Forschungsstelle Hannover (Leibniz-Archiv), whose help was invaluable to me. I also thank Anne May, director of the Gottfried Wilhelm Leibniz Bibliothek (GWLB) / Niedersächsische Landesbibliothek Hannover, as well as the staff of the GWLB, for all manner of support, in particular Rebecca Engelhardt and Jutta Wollenerg for preparing the material for the images. I thank Nora Gädeke for inspiring ideas, Siegmund Probst and Charlotte Wahl for important suggestions on the passages having to do with mathematics, Thomas Stockinger for suggested reading on early modern Vienna, Jürgen Gottschalk and Friedrich-Wilhelm Wellmer for information on the Harz, and Renate Essi and Malte-Ludolf Babin for kindly granting me permission to use the online transcriptions from Leibniz's later years. My thanks to Susanne Kupsch for insightful edits.

ABBREVIATIONS

A   *Akademieausgabe* or Academy Edition of Leibniz's work, followed by the series (*Reihe*) number (given as a Roman numeral) and the volume (*Band*) number (given as an Arabic numeral): Gottfried Wilhelm Leibniz, *Sämtliche Briefe und Schriften*, edited by Preußische Akademie der Wissenschaften (Berlin, 1993). (Today the *Akademieausgabe* is edited by the Berlin-Brandenburgischen Akademie der Wissenschaften and the Akademie der Wissenschaften zu Göttingen.) Many of the volumes have been digitized and are available for free on the website of the Gottfried Wilhelm Leibniz Gesellschaft at http://leibnizedition.de. Many other digital resources are available at this site as well, such as the table of contents spanning the entire *Akademieausgabe* and a database of persons and of the correspondence.

GM   Gottfried Wilhelm Leibniz, *Leibnizens mathematische Schriften*, edited by Carl Immanuel Gerhardt, 7 vols. (Halle, 1849–63).

GP   Gottfried Wilhelm Leibniz, *Die philosophische Schriften*, edited by Carl Immanuel Gerhardt, 7 vols. (Berlin, 1875–90).

GWLB   Gottfried Wilhelm Leibniz Bibliothek— Niedersächsische Landesbibliothek

LBr.   Leibniz correspondence, GWLB

LH   Leibniz manuscripts, GWLB

LK-MOW   Leibniz correspondence—Memory of the World, GWLB

T   Transcriptions of the Leibniz correspondence from

1708 to 1716, followed by the year of the letter cited:
Gottfried Wilhelm Leibniz, *Sämtliche Briefe und
Schriften*, revised and with new entries added by
Malte-Ludolf Babin and Renate Essi, Leibniz-Archiv
Hannover, available online: https://www.gwlb.de/leibniz
/digitale-ressourcen/repositorium-des-leibniz-archivs

GLOSSARY OF NAMES

Adelaide of Italy (931/32–999): Wife of Otto I (Otto the Great); canonized in 1097

Alencé, Joachim d' (c. 1640–1707): French astronomer, physicist, secretary to Louis XIV

Alvensleben, Carl August von (1661–97): Hofrat under Ernest Augustus, canon in Magdeburg

Ancillon, Charles (1659–1715): Franco-German jurist, diplomat, police director in Berlin

Anne Stuart (1665–1714): From 1707, Queen of Great Britain

Anthony Ulrich, Duke of Brunswick (Anton Ulrich von Braunschweig-Lüneburg) (1633–1714): Duke of Wolfenbüttel

Ariosti, Attilio (1666–1729): Italian deacon, composer

Aristotle (384–322 B.C.): Greek philosopher and universal scholar

Arnauld, Antoine (1612–94): French theologian, philosopher, mathematician

Ausson de Villarnoux, François d' (d. after 1713): Chief equerry, chamberlain to Queen Sophie Charlotte

Barclay, John (1582–1621): Scottish poet

Bayle, Pierre (1647–1706): French philosopher, writer

Becher, Johann Joachim (1635–82): German scholar, mercantilist

Behrens, Conrad Barthold (1660–1736): German doctor, historian

Benedict III (d. 858): From 855, pope

Berkeley, George (1685–1753): Irish theologian, Enlightenment philosopher

Bernoulli, Jacob (1655–1705): Swiss mathematician

Bernoulli, Johann (1667–1748): Swiss mathematician, doctor

Bernstorff, Andreas Gottlieb von (1649–1726): Premierminister of the Electorate of Hanover

Bignon, Jean-Paul (1662–1743): member of the Académie française

Binswanger, Ludwig (1881–1966): Swiss psychoanalyst

Boethius, Anicius Manlius Severinus (c. 480/85–c. 524/26): Roman scholar, theologian

Boineburg, Philipp Wilhelm von (1656–1717): took an educational trip to Paris in 1672 under Leibniz' supervision, later governor of Erfurt serving under the Elector-Archbishop of Mainz

Boole, George (1815–64): English mathematician, philosopher

Bourguet, Louis (1678–1742): Swiss polymath

Bouvet, Joachim (1656–1730): French Jesuit, missionary to China

Brandshagen, Jobst Dietrich (1659–1716): German mining expert, worked for Leibniz as a copyist and assistant on his mining projects in the Harz

Brosseau, Christophe (1630–1717): Hanoverian envoy to France

Burnett of Kemney, Thomas (1656–1729): English lawyer, traveler

Caesar, Gaius Julius (100–44 B.C.): Roman general, emperor, writer

Campanella, Tommaso (1568–1639): Italian philosopher

Carcavy, Pierre de; *also* **Carcavi (c. 1605–84):** French mathematician, librarian

Caroline of Brandenburg-Ansbach (1683–1737): wife of George Augustus, Princess of Wales; from 1727, Queen of Great Britain and Ireland

Carus, Carl Gustav (1789–1869): German polymath

Cassini, Giovanni Domenico; *also* **Gian Domenico C. or Jean-Dominique C. (1625–1712):** Franco-Italian astronomer, mathematician

Cavalieri, Bonaventura Francesco (1598–1647): Italian mathematician, astronomer

Caze, César (1641–1720): Franco-Dutch mathematician, technician

Chamberlayne, John (c. 1668–1723): English writer, translator

Charles VI (1685–1740): From 1711, Holy Roman Emperor

Charles XII (1682–1718): From 1697, King of Sweden

Charlotte Felicitas of Braunschweig-Lüneburg (1671–1710): From 1696, wife of Rinaldo d'Este

Clarke, Samuel (1675–1729): English theologian, philosopher

Colbert, Jean-Baptiste (1619–83): French statesman, finance minister

Dangicourt, Pierre (1664–1727): French mathematician, Huguenot; fled to Berlin

Darwin, Charles Robert (1809–82): British naturalist

Des Bosses, Bartholomäus (1668–1738): Belgian Jesuit, theologian, historiographer

Descartes, René (1596–1650): French philosopher, mathematician, naturalist

Des Vignoles, Alphonse (1649–1744): French clergyman, scientist

Dionysius of Halicarnassus (c. 54 B.C.–7 B.C.): Greek historian

Du Mont, Andreas (d. 1697): Generalfeldzeugmeister in Hanoverian service

Eckhart, Johann Georg von (1674–1730): German historian, librarian, Leibniz's secretary

Eichholtz, Petrus (d. 1665): Pastor in Zellerfeld

Elisabeth Charlotte; *also* **Liselotte von der Pfalz (1652–1722):** From 1671, Duchess of Orléans

Erasmus of Rotterdam, Desiderius (c. 1467–1536): Dutch scholar, philologist

Ernest Augustus, Elector of Hanover (1629–98)

Ernst von Hessen-Rheinfels, Landgrave (1623–93)

Erskine, Robert (1677–1718): Scottish doctor

Eugene of Savoy, *also* **Prince Eugene (1663–1736):** Troop commander

Fabri, Honoré (c. 1608–88): French mathematician, astronomer, philosopher, physicist

Fabricius, Johann (1644–1729): German theologian, university lecturer, consistorial counselor

Falaiseau, Pierre de (1649–1726): Brandenburgian envoy to London (1682–85)

Fardella, Michel Angelo (1650–1711): Italian astronomer, mathematician

Fechner, Gustav Theodor (1801–87): German doctor, physicist, and natural philosopher

Federl, Georg (c. 16th c.): Viennese merchant, owner of the Großer Federlhof

Fénelon, François (1651–1715): French writer; from 1695, Archbishop of Cambrai

Flemming, Jakob Heinrich von; *also* **Jacob (1667–1728):** Minister and head of the army under Augustus the Strong of Saxony

Fontaney, Jean de (1643–1710): French Jesuit, mathematician, missionary in China

Fontenelle, Bernard Le Bovier de (1657–1757): French Enlightenment philosopher, writer

Foucher, Simon (1644–96): French scholar

Fountaine, Andrew (1676–1753): English antiquarian, art collector

Frederick I (1657–1713): From 1701, the first king of Prussia

Freiesleben, Christian (1613–80): Priory administrator at the University of Leipzig

George I (Georg Ludwig von Braunschweig-Lüneburg) (1660–1727): Elector of Hanover; from 1714, King of Great Britain and Ireland

George II (Georg August von Braunschweig-Lüneburg) (1683–1760): Prince-Elector of Hanover, later Prince of Wales; from 1727, King of Great Britain and Ireland

Gödel, Kurt (1906–78): Austrian-American mathematician, logician

Gottsched, Johann Christoph (1700–66): German writer, linguist, literary theorist

Grimaldi, Claudio Filippo (1638–1712): Italian Jesuit, favorite of the Kangxi emperor (1662–1722)

Grünbein, Durs (b. 1962): German poet, professor of poetry

Guidi, Giuseppe (c. 1689–1720): Italian abbot, court poet, and operator of a news agency

Habbaeus von Lichtenstern, Christian (c. 1627–80): Swedish-Danish, served as a diplomat in Frankfurt, later in Hamburg

Hartsinck, Pieter; *also* Hartzing (c. 1633/37–80): Hofrat and Bergrat under Ernest Augustus

Hegel, Georg Wilhelm Friedrich (1770–1831): Philosopher

Helmont, Franciscus Mercurius van (c. 1614/18–99): Flemish polymath, doctor

Heraclitus of Ephesus (c. 520–c. 460 B.C.): Greek philosopher

Hermann, Jacob (1678–1733): Swiss mathematician

Hodann, Johann Friedrich (1674–1745): German theologian, pedagogue, Leibniz's secretary

Hofmann, Karl Gottlieb (1762–99): German publisher

Hooke, Robert (1635–1703): English polymath, architect

Hugony, Charles (c. 1670–c. 1714): Former *procureur au parlement* in Toulouse; in 1699 fled to Berlin, where thanks to Leibniz's recommendation he became an officer in the Prussian guard

Huldenberg, Daniel Erasmus von (1660–1733): German geographer, Hanoverian envoy to Vienna

Humboldt, Alexander von (1769–1859): German explorer, naturalist

Huygens, Christiaan (1629–95): Dutch mathematician, astronomer, physicist

Hyde, Edward, Viscount Cornbury, Third Earl of Clarendon (1661–1723): English politician, governor of New York, Queen Anne's envoy to Hanover

Innocent XI (1611–1689): From 1676, pope

Joanna Papissa; *also* Pope Joan: Fictional figure

John Frederick, Duke of Brunswick (Johann Friedrich von Braunschweig-Lüneburg) (1625–79): Duke of Brunswick-Lüneburg

Joseph I (1678–1711): From 1705, Holy Roman emperor

Kant, Immanuel (1724–1804): German philosopher

Karl Moritz, Raugraf zu Pfalz (1670–1702)

Ker, John, of Kersland (1673–1726): Scottish diplomat and spy

Kirch, Gottfried (1639–1710): Calendar maker, astronomer

Klencke, Charlotte Elisabeth von (1685–1748): Courtesan

Kniphausen, Friedrich Ernst von (1678–1731): Prussian chamberlain under Frederick I, envoy to various European courts

Knoche, Johann Barthold (c. 1682–c. 1709): Leibniz's servant

Leeuwenhoek, Antoni van (1632–1723): Dutch micrologist, naturalist

Le Gobien, Charles (1652–1708): French writer, Jesuit

Leibnütz, Friedrich (1597–1652): German notary, university instructor, father of G. W. Leibniz

Leibnütz, Johann Friedrich (1632–96): German pedagogue, professor in Leipzig, half brother of G. W. Leibniz

Leo IV (c. 790–855): From 847, pope

Leopold I (1640–1705): From 1658, Holy Roman emperor

L'Hôpital, Guillaume François Antoine de (1661–1704): French mathematician

Linsen, Hans (d. 1698): Miller, carpenter

Livy (59 B.C.–A.D. 17): Roman historian

Locke, John (1632–1704): English doctor, philosopher

Louis XIV (1636–1715): From 1643, King of France

Lucretia (d. late 6th c. B.C.): Wife of Lucius Tarquinius Collatinus

Lüde, Carol von (c. 17th c.): War secretary in Hanover

Ludolf, Hiob (1624–1704): German orientalist

Lully, Jean-Baptiste; *also* Giovanni Battista L. (1632–87): Franco-
Italian composer, violinist
Luther, Martin (1438–1546): German theologian
Luxemburg, Rosa (1871–1919): Polish-German Marxist,
socialist, Communist
Magliabechi, Antonio (1633–1714): Italian scholar, librarian
Maillet, Benoît de (1656–1738): French consul in Egypt,
traveler, geologist
Marchetti, Angelo (1674–1753): Italian astronomer, mathematician
Molanus, Gerhard Wolter; *also* G. W. van der Muelen (1633–1722):
Protestant theologian, abbot
Molière (Jean-Baptiste Poquelin) (1622–73): French actor,
playwright
Muratori, Ludovico Antonio (1672–1750): Italian
scholar, historiographer
Naudé, Philipp(e) (1654–1729): Franco-German mathematician,
Huguenot; fled to Berlin
Newton, Isaac (1643–1727): English physicist,
astronomer, mathematician
Nicholas I (820–67): From 858, pope
Nicolai, Friedrich (1733–1811): German publisher, writer
Novalis (Friedrich von Hardenberg) (1772–1801): German writer
Oldenburg, Henry (1618–77): German diplomat, natural philosopher,
secretary of the Royal Society
Orban, Ferdinand (1655–1732): German mathematician, Jesuit, court
preacher, collector of art and natural historical artifacts
Origen (185–253/54): Christian theologian and scholar
Otto the Great (912–73): From 962, Holy Roman emperor
Overbeck, Adolph-Theobald (1672–1719): German librarian, deputy
rector in Wolfenbüttel
Ovid (43 B.C.–A.D. 17): Roman poet
Paracelsus (Theophrastus Bombast von Hohenheim) (1493/94–1541):
Swiss doctor
Pascal, Blaise (1623–62): French mathematician,
physicist, philosopher
Perez, Antonio (1599–1649): Jesuit, theologian of late
Spanish Scholasticism
Peter the Great (1672–1725): From 1682, Czar of Russia

Petersen, Johann Wilhelm (1649–1727): German theologian, mystic

Pétis de la Croix, François (1622–95): French orientalist

Pfeffer, Reinhart (c. 17th c.): Lieutenant, blacksmith in Zellerfeld

Philippe I, Duke of Orléans (1640–1701)

Platen, Franz Ernst von (1631–1709): German royal tutor, General-postmeister, Reichsgraf, Geheimrat in Hanover

Plato (428/427–348/347 B.C.): Greek philosopher

Plotinus (205–70): Roman philosopher

Poelwyck, Balthasar van: Dutch merchant

Polonus, Martinus; *also* Martin von Troppau (c. 1220–78): Dominican, chronicler

Pope, Alexander (1688–1744): English poet

Pyrrho of Elis (c. 362 B.C.–c. 270 B.C.): Greek philosopher

Reiche, Jobst Christoph (1656/57–1740): German Justizrat, Kammersekretär in Hanover

Reimmann, Jakob Friedrich (1668–1743): German Protestant theologian, historian, philosopher

Rémond, Nicolas-François (1638–1725): French scholar, lead counselor of the Duke of Orléans

Rieux, Alexandre de, Marquis de Sourdéac (d. 1695): French stage and theater technician

Rinaldo d'Este (1655–1737): From 1694, Rinaldo III, Duke of Modena

Roberval, Gilles Personne de (1602–75): French mathematician

Rudolph Augustus, Duke of Brunswick-Lüneburg (Rudolf August Herzog zu Braunschweig und Lüneburg) (1627–1704): Prince of Brunswick-Wolfenbüttel

Saint-Vincent, Grégoire de (1584–1667): Flemish mathematician

Sancto Josepho, Augustinus Thomas a (1646–1717): Hungarian mathematician, Piarist

Saujon, Madame (17th c.): Owner of a Parisian boardinghouse

Scheits, Andreas (c. 1655–1735): German court painter

Schelling, Friedrich Wilhelm (1775–1854): German philosopher

Scheuchzer, Johann Jacob (1672–1733): Swiss naturalist, doctor

Schlitz, Friedrich Wilhelm von, genannt von Görtz (1647–1728): Geheimrat, Kammerpräsident in Hanover

Schönborn, Johann Philipp von (1605–73): From 1647, Elector-Archbishop of Mainz

Schopenhauer, Arthur (1788–1860): German philosopher

Schotanus van Steringa, Bernhard; *also* **Bernardus Schotanus à Sterringa (1640–1704):** Dutch physician, cartographer, surveyor

Schöttel, Joseph (d. 1760): Jurist; from 1729, imperial archivist for the court war chancellery in Vienna; amateur mathematician

Schöttel, Theobald (d. 1720): Imperial doorkeeper under Charles VI in Vienna, amateur mathematician

Schrader, Justus (1646–1720): German doctor from Helmstedt

Schubert, Gotthilf Heinrich von (1780–1860): German doctor, naturalist

Schubert, Johann David (1761–1822): German painter, copper engraver

Schulenburg, Matthias Johann von der (1661–1747): German field marshal

Scudéry, Madeleine de (1607–1701): French writer, salonnière

Seeländer, Nicolaus (1682–1744): German copper engraver, numismatist

Seip, Johann Philipp (1686–1757): German doctor

Sextus Tarquinius (d. c. 499/96 B.C.): Son of Tarquinius Superbus

Shakespeare, William (1564–1616): English playwright

Sinzendorf, Philipp Ludwig Wenzel von (1671–1742): Austrian diplomat, Hofkanzler

Sloane, Hans (1660–1753): Irish scientist, physician, botanist

Sophie Charlotte of Hanover (1668–1705): From 1688, Queen of Prussia

Sophie of Hanover (Sophie von der Pfalz) (1630–1714): Electress of Hanover, wife of Ernest Augustus, mother of Sophie Charlotte

Spedazzi, Giuseppi (17th/18th c.): Abbot, secretary in the decryption office of the imperial court in Vienna

Spinoza, Baruch de (1632–77): Dutch philosopher

Stahl, Georg Ernst (1659–1734): German chemist, physician

Standen, Anthony (c. 1548–c. 1615): English spy in the service of Mary, Queen of Scots and Anne of Denmark

Stensen, Niels (Nicolaus Steno) (1638–86): Danish physician, naturalist, bishop

St. John, Henry, Viscount Bolingbroke (1678–1751): British politician, philosopher

Sully, Henry (1680–1729): English clockmaker, worked in France with Julien Le Roy

Süßmilch, Johann Peter (1707–67): German statistician, demographer

Swammerdam, Jan (1637–80): Dutch anatomist, naturalist

Tarquinius Superbus, Lucius (d. c. 495 B.C.): Seventh and last king of Rome, according to Roman tradition; cf. Sextus Tarquinius

Tentzel, Wilhelm Ernst (1659–1707): German scholar, archivist, numismatist

Teuber, Gottfried (1656–1731): Studied mathematics and theology in Jena, confidant of Erhard Weigel

Toland, John (1670–1722): Irish philosopher, Enlightenment figure

Troyel, Isaac (b. 1686): Dutch publisher and bookseller

Tschirnhaus, Ehrenfried Walther von (1651–1708): German mathematician, naturalist

Turing, Alan (1912–54): British mathematician, computer scientist

Uffenbach, Johann Friedrich von (1687–1769): Mayor of Frankfurt am Main, patron, brother of Zacharias Konrad von Uffenbach

Uffenbach, Zacharias Konrad von (1683–1734): Writer of a travelogue, book collector

Vaget, Augustin (Augustinus Vagetius) (1670–1700): Professor of mathematics in Gießen

Valla, Lorenzo (Laurentius V) (1407–57): Italian humanist, philologist

Vallisneri, Antonio; *also* **Vallisnieri (1661–1730):** Italian physician, biologist, geologist

Vogler, Johann Hermann: Leibniz's assistant/secretary

Voltaire (François-Marie Arouet) (1694–1778): French philosopher, writer

Vota, Carlo Maurizio (1629–1715): Italian Jesuit, scholar, Father Confessor to European rulers

Wagner, Rudolf Christian (1671–1741): German mathematician, physicist

Wallenstein (1583–1634): Bohemian general

Weigel, Erhard (1625–99): Mathematician, astronomer, inventor

Wilhelmine Amalie of Brunswick (1673–1742): Consort of Emperor Joseph I; dowager empress

Wintzingerode, Hedwig Katharina von: Lady-in-waiting to Electress Sophie of Hanover

Witsen, Nicolaas (1641–1717): Diplomat, mayor of Amsterdam, geographer, cartographer

Wittgenstein, Ludwig (1889–1951): Austrian-British philosopher

Zuse, Konrad (1910–95): German computing pioneer

NOTES

Introduction

1 Voltaire, *Candide*, trans. John Butt (Harmondsworth, Middlesex: Penguin Books, 1947), 37.

2 Jakob Friedrich Reimmann, *Jacob Friederich Reimmanns, Weyland Hochverdienten Superintendentens der Evangelischen Kirchen . . . Eigene Lebens-Beschreibung oder Historische Nachricht von Sich Selbst* (Brunswick, 1745), 44.

3 Gottfried Wilhelm Leibniz, "Damit Nägel sich nicht leicht aus dem holze ziehen" (Incipit), LH 38, Bl. 147, in Ernst Gerland, *Leibnizens nachgelassene Schriften physikalischen, mechanischen, und technischen Inhalts* (Leipzig: B.G. Teubner, 1906), 244.

4 Sven K. Knebel, "Necessitas moralis ad optimum. Zum historischen Hintergrund der Wahl der besten aller möglichen Welten," *Studia Leibnitiana* 23, no. 1 (1991): 21.

5 "Ces meditations me rendoient melancolique": Leibniz, "Songe philosophique de Leibniz, parmi les Collectanea de scientia generali de Guil. Pacidius, c. 1693," handwritten superscript by Johann Daniel Gruber, LH 4, 8, Bl. 51; published in Eduard Bodemann, *Die Leibniz-Handschriften der Königlichen Öffentlichen Bibliothek zu Hannover* (Hanover: Hahn, 1895), 108.

Chapter 1: PARIS, OCTOBER 29, 1675

1 Cf. Christian Pfister and Walter Bareiss, "The Climate in Paris between 1675 and 1715 According to the Meteorological Journal of Louis Morin," *Climatic Trends and Anomalies in Europe 1675–1715. High Resolution Spatio-Temporal Reconstructions from Direct Meteorological Observations and Proxy Data: Methods and Results*, ed. Burkhard Frenzel et al. (Stuttgart: G. Fischer, 1994), 151–72.

2 Leibniz, "Imago Leibnitii," in Kurt Müller and Gisela Krönert, eds., *Leben*

und Werk von Gottfried Wilhelm Leibniz: Eine Chronik (Frankfurt: Kloster-
mann, 1969), 1–2. Cf. Georg Wilhelm Böhmer, "Leibnizens Bild von ihm
selbst entworfen," *Magazin für das Kirchenrecht, die Kirchen-und Gelehrten-
Geschichte nebst Beiträgen zur Menschenkenntniß überhaupt* (1787), 1:104–7.
For Leibniz's culinary preferences, see Herma Kliege-Biller, "Kirschsaft,
Wein und Milchkaffee. Annotatiunculae culinariae," in *Individuation,
Sýmpnoia pánta, Harmonia, Emanation: Festgabe für Heinrich Schepers zu sei-
nem 75. Geburtstag,* ed. Klaus D. Dutz (Münster: Nodus, 2000), 157–72.

3   Leibniz, "[Selbstschilderung. G.G.L.]," in *Leibnizens geschichtliche Aufsätze und Gedichte aus den Handschriften der Königlichen Bibliothek zu Hannover,* ed. Georg Heinrich Pertz (Hanover: Im Verlage der Hahnschen
Hofbuchhandlung, 1847), 4:173.

4   Cf. Michael Kempe, "Leibniz' Vorstellungen von europäischer und globaler
Politik: Der ägyptische Plan revisited," in *Leibniz in Mainz: Europäische
Dimensionen der Mainzer Wirkungsperiode,* ed. Irene Dingel et al. (Göttin-
gen: Vandenhoeck & Ruprecht, 2019), 73–91; and Benjamin Steiner, "Leib-
niz, Colbert und Afrika. Wissen und Nicht-Wissen über die geopolitische
Bedeutung des Kontinents um 1670," in *Wissenskulturen in der Leibniz-Zeit.
Konzepte—Praktiken—Vermittlung,* ed. Friedrich Beiderbeck and Claire
Gantet (Berlin: De Gruyter Oldenbourg, 2021), 75–112.

5   Leibniz, "Imago Leibnitii," 1.

6   Cf. A. Roger Ekirch, *At Day's Close: Night in Times Past* (New York:
Norton, 2005).

7   Leibniz to Simon Foucher, 1675. A II, 1 N. 120, 386–92.

8   Leibniz, "scripta brevia feminae sine manibus," Paris, March 1674, A VIII,
2 N. 79, 696.

9   Leibniz to Duke John Frederick, January 21, 1675, A I, 1 N. 328, 491–93:
"Paris est un lieu, ou il est difficile de se distinguer: on y trouue les plus
habiles hommes du temps, en toutes sortes des sciences, et il faut beaucoup
de travail, et un peu de solidité, pour y establir sa reputation."

10   Leibniz, "Imago Leibnitii," 1.

11   Leibniz, "Drôle de Pensée, touchant une nouuelle sorte de représentations,"
A IV, 1 N. 49, 562–68.

12   Leibniz to Jean Gallois, November 2, 1675, A III, 1 N. 67, 304.

13   Leibniz, [no date], LH 41, 10 Bl. 2: "il me vient quelques fois tant de pensées
le matin dans une heure, pendant que je suis encor au lit, que j'ay besois
d'employer toute la matinée et par fois toute la journée et au de là, pour les
mettre distinctement par ecrit." Cited in Eduard Bodemann, *Die Leibniz-
Handschriften der Königlichen Öffentlichen Bibliothek zu Hannover 1895*
(Hanover: Hahn, 1895), 338.

14   Cf. Rudolf Christian Wagner to Leibniz, June 26, 1700, A III, 8 N. 172,

443. For the records of the conversations between Leibniz and Tschirn-
haus in Paris, cf. Uwe Mayer, "Mündliche Kommunikation und schriftliche
Überlieferung—Die Gesprächsaufzeichnungen von Leibniz und Tschirn-
haus zur Infinitesimalrechnung aus der Pariser Zeit," in *VIII. Internationaler
Leibniz Kongress: Einheit in der Vielfalt. Vorträge 1. Teil*, ed. Herbert Breger,
Jürgen Herbst, und Sven Erdner (Hanover, 2006), 588–94.

15   Leibniz, "Rechnungen zur Einkaufsliste," A VIII, 2 N. 80, 697: "2 saucisses
2 poulets 4 pain. 3 chopin."

16   Leibniz to Rudolf Christian von Bodenhausen, November 26 / December 6,
1697, A III, 7 N. 162, 652: "ein groß chaos."

17   Hans Magnus Enzensberger, "G.W.L. (1646–1716)," in Enzensberger,
Mausoleum: Siebenunddreißig Balladen aus der Geschichte des Fortschritts
(Frankfurt: Suhrkamp, 1975), 24.

18   Cf. Leibniz, "Imago Leibnitii," 1.

19   Cf. Jan Lazardzig, *Theatermaschine und Festungsbau: Paradoxien der Wissensproduktion im 17. Jahrhundert* (Berlin: Akademie Verlag, 2007), 230.

20   Leibniz to Christian Habbaeus von Lichtenstern, May 5, 1673, A I, 1 N.
277, 416: "n'estant pas accoûtumé à m'assujettir à de certaines caprices poli
tiqves de qvelqves Grands Seigneurs et aimant mieux d'estre éloigné de
toutes ces affaires la, qve de viure dans des inqvietudes continuelles."

21   Leibniz, draft of a letter to Peter I [January 16, 1712], cited in Woldemar
Guerrier, *Leibniz in seinen Beziehungen zu Russland und Peter dem Grossen:
Eine geschichtliche Darstellung dieses Verhältnisses nebst den darauf bezüglichen
Briefen und Denkschriften* (St. Petersburg, 1873), 208.

22   Leibniz to Christian Freiesleben, October 11/21, 1675, A I, 1 N. 287, 427–
31 (the draft is dated October 20 in Leibniz's hand); Leibniz to Johann
Friedrich Leibniz, October 11/21, 1675, A I, 1 N. 288, 431–33.

23   Joachim d'Alencé to Leibniz, [October 29, 1675], A III, 1 N. 66, 302–3.

24   "Pour moy je croy qve je seray un Amphibie, tantost en Allemagne, tantost
en France." Leibniz to Christian Habbaeus von Lichtenstern, February 14,
1676, A I, 1 N. 298, 445.

25   Cf. Johann Friedrich Leibniz to Leibniz, May 6, 1675, A I, 1 N. 284, 423–24.

26   Cited in Kathleen M. Lea, "Sir Anthony Standen and Some Anglo-Italian
Letters," *English History Review* 47 (1932): 477.

27   Leibniz to Johann Friedrich Leibniz, October 11/21, 1675, A I, 1,
N. 288, 432.

28   Leibniz for Philipp Wilhelm von Boineburg, c. March 1673, A I, 1 N.
226, 332–33.

29   Leibniz, "Calculus per divisiones loco multiplicationum," October 29, 1675,
LH 35, 4, Bl. 19.

30   Cf. Eberhard Knobloch, "Die entscheidende Abhandlung von Leibniz zur

Theorie linearer Gleichungssysteme," *Studia Leibnitiana* 4 (1972): 163–80; and Achim Trunk, "Sechs Systeme. Leibniz und seine *signa ambigua*," in *"Für unser Glück oder das Glück anderer." Vorträge des X. Internationalen Leibniz-Kongresses, Hannover, 18.-23. Juli 2016*, ed. Wenchao Li and Ute Beckmann, 4:191–207 (Hildesheim: Georg Olms Verlag, 2016). The Leibniz pieces on double and composite signs can be found in A VII, 7, in particular N. 2, 8, 11–15 and 45–46.

31   Davide Crippa, *The Impossibility of Squaring the Circle in the 17th Century: A Debate Among Gregory, Huygens and Leibniz* (Cham: Birkhäuser, 2019).

32   Cf. Herbert Breger, "Die mathematisch-physikalische Schönheit bei Leibniz," in *Kontinuum, Analysis, Informales—Beiträge zur Mathematik und Philosophie von Leibniz*, ed. Wenchao Li (Berlin: Springer, 2016), 111. For a critical view of the role of aesthetics in mathematics and the natural sciences, see also Sabine Hossenfelder, *Das hässliche Universum: Warum unsere Suche nach Schönheit die Physik in die Sackgasse führt* (Frankfurt: Fischer, 2018).

33   Cf. Ohad Nachtomy, "A Tale of Two Thinkers, One Meeting, and Three Degrees of Infinity: Leibniz and Spinoza (1675–78)," *British Journal for the History of Philosophy* 19 (2011): 935–61.

34   "Cum Deus calculat et cogitationem exercet fit mundus." Leibniz, "Dialogus," August 1677, A VI, 4, Part A N. 8, 22.

35   Leibniz, "Additio ad schedam in Actis proxime antecedentis Maji pag. 233 editam, de dimensionibus curvilineorum," per G.G.L., *Acta eruditorum* (December 1684): 585–87. German translation: Leibniz, "Ergänzung zum Beitrag *Wie man zur Ausmessung von Figuren kommt*, der in den *Acta* des letztvergangenen Mai auf S. 233 f. veröffentlicht ist," *Die mathematischen Zeitschriftenartikel*, trans. Heinz-Jürgen Heß and Malte-Ludolf Babin (Hildesheim: Georg Olms, 2011), 47–48.

36   "Utile erit scribi ∫ pro omn": Leibniz, "Analyseos Tetragonisticae pars secunda," Paris, October 29, 1675, A VII, 5 N. 40, 292.

37   Leibniz, "Methodi tangentium inversae exempla," Paris, November 11, 1675, A VII, 5 N. 46, 324.

38   Cf. Herbert Breger, "Leibniz's Calculation with Compendia," in *Kontinuum, Analysis, Informales—Beiträge zur Mathematik und Philosophie von Leibniz*, ed. Wenchao Li (Berlin: Springer, 2016), 147–58.

39   Cf. Siegmund Probst, "The Calculus," in *The Oxford Handbook of Leibniz*, ed. Maria Rosa Antognazza (Oxford: Oxford Academic, 2018), 213–15, as well as Charlotte Wahl, "Assessing Mathematical Progress: Contemporary Views on the Merits of Leibniz's Infinitesimal Calculus," in *Natur und Subjekt. Vorträge des IX. Internationalen Leibniz-Kongresses, 3. Teil*, ed. Herbert Breger et al. (Hanover: Gottfried-Wilhelm-Leibniz-Gesellschaft, 2011), 1172–82.

40 Cf. Alain Corbin, *The Foul and the Fragrant: Odor and the French Social Imagination* (Cambridge, Mass.: Harvard University Press, 1986).

41 Leibniz, "Imago Leibnitii," 1.

42 Cf. Sybille Krämer, *Berechenbare Vernunft: Kalkül und Rationalismus im 17. Jahrhundert* (Berlin: Walter de Gruyter, 1991), 154–58.

43 Leibniz, "Dissertatio de arte combinatoria," 1666, A VI, 1 N. 8, 163–230.

44 Leibniz to Henry [Heinrich] Oldenburg, December 28, 1675, A III, 1 N. 70, 331.

45 Leibniz to Philip Jakob Spener, July 8/18, 1687, A II, 2 N. 48, 213.

46 Leibniz, "De numeris characteristicis ad linguam universalem constituendam," c. Spring or summer 1679, A VI, 4, Part A N. 66, 265. For a German translation, cf. Leibniz, "Anfangsgründe einer allgemeinen Charakteristik" (1677), *Schriften zur Logik und zur philosophischen Grundlegung von Mathematik und Naturwissenschaft*, vol. 4 of *Philosophische Schriften*, trans. and ed. Herbert Herring (Frankfurt: Insel Verlag, 1992), 47.

Chapter 2: ZELLERFELD (IN THE HARZ MOUNTAINS), FEBRUARY 11, 1686

1 Leibniz, "Münzwert [1686]," A IV, 3 N. 46, 406.

2 Cf. Rüdiger Glaser, *Klimageschichte Mitteleuropas: 1200 Jahre Wetter, Klima, Katastrophen*, 2nd ed. (Darmstadt: Wissenschaftliche Buchgemeinschaft, 2008), 169.

3 Johann Georg von Eckhart, "Lebensbeschreibung des Freyherrn von Leibnitz," in *Journal zur Kunstgeschichte und zur allgemeinen Literatur, 7. Theil*, ed. Christoph Gottlieb von Murr (Nürnberg, 1779), 123–231; reprinted in J. A. Eberhard and J. G. Eckhart, eds., *Leibniz-Biographien* (Hildesheim: George Olms, 1982), 196.

4 Anton Heinrich Meyer, "Paßzettel," November 23 / December 3, 1680, A I, 3 N. 71, 103.

5 Between 1683 and 1686, Leibniz also stays at times at the home of the mining factor Jobst Philipp Rochoff. Likewise located on present-day Bornhardtstraße is the home of the town theologian and superintendent Caspar Calvör, whom Leibniz visits often in this period. My thanks to Jürgen Gottschalk and Friedrich-Wilhelm Wellmer for information on this front.

6 For the aural environment of early modern cities, cf. David Garrioch, "Sounds of the City: The Soundscape of Early Modern European Towns," *Urban History* 30 (May 2003): 5–25.

7 Leibniz to Hermann Conring, [June 1678], A II, 1 N. 181, 633.

8   Cf. Vincenzo De Risi, *Geometry and Monadology: Leibniz's "Analysis Situs"* *and Philosophy of Space* (Basel: Birkhäuser, 2007).

9   For his own experiments with underwater vessels, cf. Leibniz, "De navi submarina," May 1678, LH 38, 103.

10 Cf. Gerda Utermöhlen, "Leibnizens Hundebittschrift, der Papinsche Topf und die Geschichte eines hartnäckigen Lesefehlers," *Hannoversche Geschichtsblätter, Neue Folge* 33 (1979): 57–62.

11 *Oxford English Dictionary*, s.v. "cameralism."

12 Cf. Jobst Dietrich Brandshagen to Leibniz, December 23, 1685 / January 2, 1686, A I, 4 N. 191, 249–50.

13 Cf. Friedrich-Wilhelm Wellmer and Jürgen Gottschalk, "Leibniz' Scheitern im Oberharzer Silberbergbau: Neu betrachtet, insbesondere unter klimatischen Gesichtspunkten," *Studia Leibnitiana* 42 (2010): 205.

14 Jobst Dietrich Brandshagen to Leibniz, December 23, 1685 / January 2, 1686, A I, 4 N. 191, 249–50.

15 For disputes around Leibniz's experiments, see, e.g., "Treibkunst des Hofrats Leibniz bei der Grube Rosenhof, 1685–1686," Bergarchiv Clausthal (Außenstelle des Niedersächsischen Landesarchivs Hannover), NLA HA, BaCl Hann. 84a, Nr. 6747.

16 For this bit of information, I am indebted to the head of the Clausthal-Zellerfeld town archives, Helge Frank.

17 Hans Linsen to Leibniz, January 30 / February 9, 1686, A III, 4 N. 133, 265–66.

18 Leibniz, "Examen religionis christianae (Systema Theologicum)," c. April–October 1686, A VI, 4 Teil C N. 420, 2355–455.

19 The paper comes from the "Untere Glockenmühle an der Grane" paper mill in the Harz (near Hahnenklee) and bears its characteristic wild man watermark alternating with the sign of a bell. Cf. Eberhard Tacke and Irmgard Tacke, "Von den Papiermachern zur Grane und ihren Wasserzeichen," ed. K. W. Sanders, in *Harz-Zeitschrift* 17 (1965): 79–104.

20 "Dieu est un estre absolument parfait," "L'idée de Dieu," "La notion de Dieu," LH 1, 3, 1 Bl. 1r.

21 Leibniz to Landgrave Ernst von Hessen-Rheinfels, January 1/11, 1684, A I, 4 N. 285, 319–22.

22 The letter: Leibniz to Landgrave Ernst von Hessen-Rheinfels, February 1/11, 1686, A II, 2 N. 1, 3–4 (previously: A I, 4 N. 334). The enclosed summary: Leibniz to Antoine Arnauld, care of Landgrave Ernst von Hessen-Rheinfels, [February 11, 1686], A II, 2 N. 2, 4–8. Abstract of the letter and draft of the enclosed summary in Leibniz's hand: LBr. 16, 46–47.

23 Leibniz, "Discours de métaphysique," c. early 1686, A VI, 4. Teil B N. 306, 1529–90.

24 Böhmer, "Leibnitzens Bild," 105.

25 Leibniz to Arnauld care of Landgrave Ernst von Hessen-Rheinfels, [February 11, 1686], A II, 2 N. 3, 8–9.

Trans.: *In translating the paraphrased passages in this section, I consulted: Leibniz, Philosophical Writings, ed. G.H.R. Parkinson, trans. Mary Morris and G.H.R. Parkinson (Dent: London, 1973); Leibniz and Antoine Arnauld, The Leibniz-Arnauld Correspondence, trans. H. T. Mason (New York: Barnes & Noble/Manchester University Press, 1967); Leibniz and Antoine Arnauld, The Correspondence between Leibniz and Arnauld, trans. Jonathan Bennett (2017), Earlymoderntexts.com.*

26 Antoine Arnauld to Landgrave Ernst von Hessen-Rheinfels, March 13, 1686, A II, 2 N. 3, 8–9.

27 Landgrave Ernst von Hessen-Rheinfels to Leibniz, March 13/23, 1685, A I, 4 N. 309, 359–61.

28 Leibniz to Landgrave Ernst von Hessen-Rheinfels for Antoine Arnauld, [April 12, 1686], A II, 2 N. 4, S. 10–21.

29 Cf. Tilman Ramelow, *Gott, Freiheit, Weltenwahl: Der Ursprung des Begriffes der besten aller möglichen Welten in der Metaphysik der Willensfreiheit zwischen Antonio Perez S. J. (1599–1649) und G. W. Leibniz (1646–1716)* (Leiden: E.J. Brill, 1997).

30 Cf. Eric Stencil, "Arnauld's God Reconsidered," *History of Philosophy Quarterly* 36 (2019): 19–38.

31 Leibniz to Landgrave Ernst von Hessen-Rheinfels for Antoine Arnauld, [February 11, 1686], A II, 2 N. 2, 5 (Thesis 7) and 7 (Thesis 30); Leibniz, *Discours de métaphysique* [early 1686], A VI, 4. Teil B N. 306, 1538–39 and 1575–76.

32 Leibniz to Landgrave Ernst von Hessen-Rheinfels, February 1/11, 1686, A II, 2 N. 2, 5 (Thesis 4); Leibniz, *Discours de métaphysique*, c. early 1686, A VI, 4. Teil B N. 306, 1535–36.

33 Leibniz to Duke Ernest Augustus, Promemoria, c. March 1684, A I, 4 N. 26, 43–44.

34 Leibniz to Landgrave Ernst von Hessen-Rheinfels for Antoine Arnauld, [February 11, 1686], A II, 2, N. 2, 5 (Thesis 5).

35 Leibniz, *Discours de métaphysique*, 1536 (Thesis 5).

36 Hans Peter Münzenmayer, "Leibniz' Inventum Memorabile: Die Konzeption einer Drehzahlregelung vom März 1686," *Studia Leibnitiana* 8 (1976): 113–19.

37 Cf. Friedrich-Wilhelm Wellmer and Jürgen Gottschalk, "Bergbau und Geologie," in *Gottfried Wilhelm Leibniz: Rezeption, Forschung, Ausblick*, ed. Friedrich Beiderbeck, Wenchao Li, and Stephan Waldhoff (Stuttgart: Franz Steiner Verlag, 2020), 777–87.

38   Cf. Joseph Vogl, "Leibniz, Kameralist," in *Europa: Kultur de Sekretäre*, ed.
     Bernhard Siegert and Joseph Vogl (Zurich: Diaphanes Verlag, 2003), 97–109.

39   Leibniz, "Brevis demonstratio erroris memorabilis Cartesii," *Acta eruditorum*
     (March 1686): 161–63.

40   Leibniz, "Von dem Verhängnisse" (1695), in *Hauptschriften zur Grundle-*
     gung der Philosophie, ed. Ernst Cassirer and Artur Buchenau (Hamburg: F.
     Meiner, 1996), 2:338.

41   "Le meilleur de tous les maistres": Leibniz, *Discours de métaphysique*, c. early
     1686, A VI, 4, Part B N. 306, 1536 (Thesis 4).

42   Leibniz for Duke Ernest Augustus (?), "Denkschrift betr. die allgemeine
     Verbesserung des Bergbaues im Harz," February 20–22 / March 2–4, 1682,
     A I, 3 N. 124, 149–67.

43   Cf. Vogl, "Leibniz, Kameralist," 107; and Sybille Krämer, "Leibniz ein
     Vordenker der Idee des Netzes und des Netzwerks?" in *Vision als Auf-*
     gabe: Das Leibniz-Universum im 21. Jahrhundert, ed. Martin Grötschel
     et al. (Berlin: Berlin-Brandenburgische Akademie der Wissenschaft,
     2016), 55.

44   Bernhard Siegert, *Passagen des Digitalen: Zeichenpraktiken der neuzeitlichen*
     Wissenschaft 1500–1900 (Berlin: Brinkmann & Bose, 2003), 163.

45   Cf. Niedersächsisches Landesarchiv, Hauptstaatsarchiv Hannover, Han-
     nover HstA Cal. Br. 4, 82, Bl. 857 and Hannover HstA Cal. Br. 4, 234,
     Bl. 31–72.

46   For more examples, cf. Wellmer and Gottschalk, "Bergbau und Geologie."

47   Cf. Jon Elster, *Leibniz et la formation de l'esprit capitaliste* (Paris: Aubier
     Montaigne, 1975).

48   Cf. Ingetrud Pape, "Von den 'möglichen Welten' zur 'Welt des Möglichen'.
     Leibniz im modernen Verständnis," in *Akten des Internationalen Leibniz-*
     Kongresses, Hannover, 14.–19. November 1966, ed. Kurt Müller and Wil-
     helm Totok, vol. 1 *Metaphysik—Monadenlehre* (Wiesbaden: F. Steiner,
     1968), 266–87.

49   Leibniz, "Schlagen von Münzen," c. 1683, A IV, 3 N. 45, 401–2.

50   Leibniz for Duke Ernest Augustus, "Proposition de commerce, première
     partie: de la manufacture du lin," c. 1680, A I, 3 N. 74, 104–5; Leibniz
     for Duke Ernest Augustus, "Seconde partie," 1680, A I, 3 N. 75, 105–8.
     For more information on Pieter Hartsinck (Hartzing), cf. Jürgen Stock and
     Rainer Weichert, *Die Hartzings: Der Aufstieg einer Moerser Familie unter der*
     Ostindischen Kompanie (Duisburg: Mercator Verlag, 2020).

51   Leibniz, "Schlagen von Münzen," 402.

52   Leibniz, "Was bey Verbesserung des Münzwesens zu beobachten," [1681],
     A IV, 3 N. 42, 394–98.

Chapter 3: HANOVER, AUGUST 13, 1696

1 Leibniz, "Tagebuch" [August 1696–April 1697], *Leibnizens geschichtliche Aufsätze und Gedichte aus den Handschriften der Königlichen Bibliothek zu Hannover*, ed. Georg Heinrich Pertz (Hanover: Im Verlage der Hahnschen Hofbuchhandlung, 1847), 183. The original manuscript is in LH 41, 4 Bl. 1–62.

2 Leibniz, "Tagebuch," entry for August 3/13, 1696.

3 Leibniz for Justus Schrader, "Aufzeichnungen über seinen Gesundheitszustand," c. June 9–11, 1695, A I, 11 N. 338, 494–97; Justus Schrader for Leibniz, "Ärztliche Diagnose und Behandlungsvorschläge," [July] 25, 1695, A I, 11 N. 401, 580.

4 Cf. Leibniz for Justus Schrader, "Aufzeichnungen über seinen Gesundheitszustand," 495; Leibniz to Conrad Barthold Behrens, [February 6, 1696], A I, 12 N. 264, 396–98; and Leibniz to Thomas Burnett of Kemney, March 7/17, 1696, A I, 12 N. 309, 476.

5 Cf. Ekkehard Göhrlich, "Leibniz als Mensch und Kranker," medical dissertation (Hanover, 1987), 149 62.

6 Thomas Burnett of Kemney to Leibniz, January 27 / February 6, 1696, A I, 12 N. 266, 400.

7 Leibniz, "Tagebuch," entry for August 3/13, 1696.

8 Antonio Magliabechi to Leibniz, July 24, 1696, A I, 12 N. 462, 716–20. Cf. also Leibniz, "Tagebuch," entry for August 3/13, 1696, 183.

9 Anonymous, *Die heutigen Christlichen Souverainen von Europa, Das ist: Ein kurtzer Genealogischer und Politischer Abriß*... (Breslau, 1704), 946. Cf. Georg Schnath, *Geschichte Hannovers im Zeitalter der neunten Kur und der englischen Sukzession 1674–1714* (Hildesheim: A. Lax, 1976), 2:384.

10 Leibniz to Thomas Burnett of Kemney, December 7/17, 1696, A I, 12 N. 309, 476.

11 Cf. Gerda Utermöhlen, "Leibniz im kulturellen Rahmen des hanoverschen Hofes," in *Leibniz und Niedersachsen: Tagung anlässlich des 350. Geburtstages von G. W. Leibniz, Wolfenbüttel, 1996*, ed. Herbert Breger and Friedrich Niewöhner (Stuttgart: Franz Steiner Verlag, 1999), 213–26.

12 Cf. Günter Scheel, "Leibniz als politischer Ratgeber des Welfenhauses," in *Leibniz und Niedersachsen*, 46.

13 Leibniz to Rudolf Christian von Bodenhausen, September 20/30, 1697, A III, 7 N. 141, 574–75.

14 Angelo Marchetti to Leibniz, June 30, 1696, A III, 6 N. 245, 807–8.

15 Leibniz to Rudolf Christian von Bodenhausen, Hanover, June 18/28, 1696, A III, 6 N. 244, 804–7.

16 Leibniz to Johann Bernoulli, June 16/26, 1696, A III, 6 N. 243, 795–803.

17   Leibniz to Augustinus Vagetius, June 5/15, 1696, A III, 6 N. 239, 781. Cf. David Rabouin, "A Fresh Look at Leibniz' *Mathesis Universalis*," in *"Für unser Glück oder das Glück anderer": Vorträge des X. Internationalen Leibniz-Kongresses*, ed. Wenchao Li and Ute Beckmann (Hildesheim: Georg Olms Verlag, 2016), 4:510.

18   Cf. Leibniz to Guillaume François Antoine de L'Hôpital, Marquis de Sainte-Mesme et du Montellier, [mid-March 1693], A III, 5 N. 138, 506.

19   Leibniz to Thomas Burnett of Kemney, March 7/17, 1696, A I, 12 N. 309, 476.

20   Johann Hermann Vogler, quoted in Paul Ritter, "Bericht eines Augenzeugen über Leibnizens Tod und Begräbnis," *Zeitschrift des Historischen Vereins für Niedersachsen* 81 (1916): 251.

21   Leibniz, "Tagebuch," entry for August 3/13, 1696, 183.

22   Leibniz to Andreas Du Mont, July 11/21, 1696, A III, 7 N. 7, 26–32.

23   Leibniz to Electress Sophie, October 31, 1705, A I, 25 N. 159, 246–47. For other sources for the leaf episode, cf. Utermöhlen, "Leibniz im kulturellen Rahmen," 223n42.

24   Elisabeth Charlotte of Orléans to Electress Sophie, June 26, 1696, in *Aus den Briefen Herzogin Elisabeth Charlotte von Orléans an die Kurfürstin Sophie von Hanover: Ein Beitrag zur Kulturgeschichte des 17. und 18. Jahrhunderts, Hannover 1891*, ed. Eduard Bodemann (Hanover: Hahn, 1891), 248.

25   "Capables de mettre en colere un Theologien ordinaire": Leibniz to Lorenz Hertel, January 8/18, 1695, A I, 11 N. 14, 18.

26   Gerhard Biller, "Einleitung," A II, 3, LIV. For more on Leibniz and Helmont, cf. Gerd van den Heuvel, "Leibniz und die Sulzbacher Protagonisten Christian Knorr von Rosenroth und Franciscus Mercurius van Helmont," *Morgen-Glantz: Zeitschrift der Christian Knorr von Rosenroth-Gesellschaft* 11 (2001): 77–104.

27   Leibniz to Thomas Burnett of Kemny, March 7/17, 1696, A I, 12 N. 309, 478.

28   Duchess Elisabeth Charlotte of Orléans to Electress Sophie, August 2, 1696, A I, 13 N. 37, 705. The basis for the letter printed here is the excerpt made by Leibniz on August 13. The letter itself is located at the State Archive of Lower Saxony in Hanover under the catalog number Hann. 91 Kurf. Sophie Nr. 1, VI Bl. 138–42.

29   Duchess Elisabeth Charlotte to Electress Sophie, August 2, 1696, 705–6.

30   Leibniz for Electress Sophie and Duchess Elisabeth Charlotte von Orléans, "Bedenken über den Brief der Herzogin Elisabeth Charlotte vom 2. August 1696," [mid-August 1696], A I, 13 N. 7, 13–14.

31   Leibniz for Electress Sophie and Duchess Elisabeth Charlotte von Orléans, "Bedenken," 12.

32 Leibniz, "Système nouveau de la nature et de la communication des sub-
    stances, aussi vien que de l'union qu'il y a entre l'âme et le corps," *Journal des
    Sçavans* (June 27, 1695): 294–300, and (July 4, 1695): 301–6. Also printed
    in GP 4:471–87 (first and second drafts). English Translation: Leibniz, *Phi-
    losophical Writings*, ed. G.H.R. Parkinson, trans. Mary Morris and G.H.R.
    Parkinson, 115–32.

33 Leibniz, "Système nouveau," GP 4:480.

34 Robert Hooke, *Micrographia: or Some Physiological Descriptions of Minute
    Bodies Made by Magnifying Glasses . . .* (London: 1667), 184–85. For sim-
    ilar examples, see also the work by Francesco Redi (which was known to
    Leibniz), *Esperienze intorno alla generazione degli insetti* (1668), reprinted in
    Florence in 1996. For suggestions on this point, I thank Alessandro Becchi.

35 Thus we read in the diary, in the entry on August 16, 1696: "Gave point by
    point notes to Electress in her cabinet at Herrenhausen on what she could
    reply to Madame re: the soul; as according to Herr Helmont. Kept a copy."
    Leibniz, "Tagebuch," entry for August 6/16, 1696, 188; Leibniz for Electress
    Sophie, "Entwurf für die Antwort der Kurfürstin Sophie auf den Brief der
    Herzogin Elisabeth Charlotte vom 2. August 1696," [August 16, 1696], A
    I, 13 N. 8, 14–15.

36 Leibniz for Electress Sophie and Duchess Elisabeth Charlotte of Orléans,
    "Stellungnahme zu den Lehren F.M. van Helmonts," A I, 13 N. 41, 46.

37 Elisabeth Charlotte of Orléans to Electress Sophie, October 30, 1696, in
    Bodemann, *Aus den Briefen*, 259. Likewise found in an excerpt made for
    Leibniz: Electress Sophie to Leibniz, [early November 1696], A I, 13 N.
    56, 80–81.

38 Ohad Nachtomy, "Infinity and Life: The Role of Infinity in Leibniz's The-
    ory of Living Beings," in *The Life Sciences in Early Modern Philosophy*, ed.
    Ohad Nachtomy and Justin E. H. Smith (Oxford: Oxford University Press,
    2014), 9–28.

39 Leibniz, "Stellungnahme zu den Lehren F.M. van Helmonts," 47.

40 Leibniz to Electress Sophie for Duchess Elisabeth Charlotte, October 28
    / November 7–November 4/14, 1696. Enclosed with N. 58 (?), A I, 13 N.
    59, 87: "Quand j'estois petit garçon je prenois plaisir à voir ressusciter des
    mouches noyées, en les ensêvelissant sous la craye reduite en poussière." Ger-
    man translation: Leibniz and Electress Sophie of Hanover, in *Briefwechsel*,
    ed. Wenchao Li, trans. Gerda Utermöhlen and Sabine Sellschopp (Göttin-
    gen: Wallstein Verlag, 2017), 182.

41 Duchess Elisabeth Charlotte to Electress Sophie, August 2, 1696, A I, 13
    N. 37, 706.

42 Duchess Elisabeth Charlotte to Electress Sophie, July 26, 1696, in Bode-
    mann, *Aus den Briefen*, N. 246, 250.

43   Leibniz, "Stellungnahme zu den Lehren F. M. van Helmonts," 49.

44   Cf. Ingo Uhlig, *Traum und Poiesis: Produktive Schlafzustände 1641–1810* (Göttingen: Wallstein Verlag, 2015), 99.

45   Leibniz, "Nouveaux essais sur l'entendement humain" [summer 1703–summer 1705], A VI, 6 N. 2, 54 (préface).

46   Ludwig Binswanger, *Wandlungen in der Auffassung und Deutung des Traumes: Von den Griechen bis zur Gegenwart* (Berlin: Springer, 1928), 6, emphasis in the original.

47   Leibniz, "Günstiger Leser," foreword to Anicius Manlius Severinus Boethius, *Consolatio Philosophiae oder Christlich-vernufft-gemesser Trost und Unterricht in Widerwertigkeit und Bestürzung über dem vermeinten Wohl- oder Übel-Stand der Bösen und Frommen . . .* , trans. Franciscus Mercurius Helmont (Lüneburg, 1697).

48   Cf. Hanns-Peter Neumann, "Monadentheorie und Monadologie," in *Gottfried Wilhelm Leibniz: Rezeption, Forschung, Ausblick*, ed. Friedrich Beiderbeck, Wenchao Li, and Stephan Waldhoff (Stuttgart: Franz Steiner Verlag, 2020), 500–7.

49   Jürgen Jost, *Leibniz und die moderne Naturwissenschaft* (Berlin: Springer, 2019), 159.

50   Leibniz to L'Hôpital, July 12/22, 1695, A III, 6 N. 149, 451.

51   Rüdiger Glaser, *Klimageschichte Mitteleuropas: 1200 Jahre Wetter, Klima, Katastrophen*, 2nd ed. (Darmstadt: Wissenschaftliche Buchgemeinschaft, 2008), 174.

52   Leibniz, "Tagebuch," entry for August 3/13, 1696, 184.

53   Cf. Maria Rosa Antognazza, *Leibniz: An Intellectual Biography* (Cambridge: Cambridge University Press, 2009), 2–3.

54   On this and what follows, cf. Dominik Perler, "Was ist eine Person? Überlegungen zu Leibniz," *Deutsche Zeitschrift für Philosophie* 64, no. 3 (2016): 329–51.

55   Leibniz to Antoine Arnauld, [July 14, 1686], A II, 2 N. 14, 67–84.

56   "Theistische[r] Apfel": Perler, "Was ist eine Person?," 350.

57   Cf. Tobias Cheung, *Die Organisation des Lebendigen: Die Entstehung des biologischen Organismusbegriffs bei Cuvier, Leibniz und Kant* (Frankfurt: Campus, 2000). Cf. also François Duchesneau, *Organisme et corps organique de Leibniz à Kant* (Paris: Librairie Philosophique J. Vrin, 2018).

58   Cf. Justin E. H. Smith, "Lebenswissenschaften," in *Gottfried Wilhelm Leibniz: Rezeption, Forschung, Ausblick*, 772–73. See also Smith, *Divine Machines: Leibniz and the Sciences of Life* (Princeton, NJ: Princeton University Press, 2011).

59   Leibniz to Wilhelm Ernst Tentzel, June 8/18, 1696, A I, 12 N. 413, 638–40; Leibniz to Wilhelm Ernst Tentzel, June 17/27, 1696, A I, 12 N. 424,

661–62; and Leibniz to Wilhelm Ernst Tentzel, July 10/20, 1696, A I, 12 N. 455, 707–8.

60   "Non assero *species* aliquas *perisse*, quanquam id absurdum esse non ausim dicere: sed distinguendum censeo inter species extinctas, et valde commutatas. Sic canis et lupus, felis et tigris possunt ejusdem peciei videri. Idem dici potest de animalibus amphibicis seu bobus marinis olim Elephanto analogis." Leibniz to Wilhelm Ernst Tentzel, August 3/13, 1696, A I, 13 N. 131, 204, emphasis in the original.

61   Leibniz to Michel Angelo Fardella, September 3/13, 1696, A II, 3 N. 72, 192. Cf. Herma Kliege-Biller, Stephan Meier-Oeser, and Stephan Waldhoff, "Einen barocken Universalgelehrten edieren: Gottfried Wilhelm Leibniz, Sämtliche Schriften und Briefe," *Deutsche Zeitschrift für Philosophie* 64, no. 6 (2016): 951–97.

Chapter 4: BERLIN, APRIL 17, 1703

1   "Ita habemus solutum aenigma quod Sinenses a non uno annorum millenario non intellexere, amissaque genuina significatione, interpretationes peregrinas sunt commenti." Leibniz to Hans Sloane, April 17, 1703, British Library, Sloane Ms 4039, pp. 116r–117. Print version: Eric J. Aiton, "An Unpublished Letter of Leibniz to Sloane," *Annals of Science* 38, no. 1 (2006): 103–7, 105. A print reproduction of the letter for inclusion in the Akademieausgabe is in preparation and planned for A III, 9.

2   Cf. Sabine Sellschopp, " 'Eine kleine tour nach Hamburg incognito': Zu Leibniz' Bemühungen von 1701 um die Position eines Reichshofrats," *Studia Leibnitiana* 37, no. 1 (2005): 68–82.

3   Cf. Gerd van den Heuvel, "Leibniz im Netz: Die frühneuzeitliche Post als Kommunikationsmedium der Gelehrtenrepublik um 1700," *Lesesaal: Kleine Spezialitäten aus der Gottfried Wilhelm Leibniz Bibliothek—Niedersächsische Landesbibliothek* 32 (2009): 19–26.

4   Cf. Lloyd Strickland, ed., *Leibniz and the Two Sophies: The Philosophical Correspondence* (Toronto: Iter, 2011).

5   Cf. Barbara Beuys, *Sophie Charlotte: Preußens erste Königin* (Berlin: Insel Verlag, 2018), 279–98.

6   Electress Sophie to Leibniz, October 4, 1702, A I, 21 N. 63, 80.

7   Leibniz to Philipp Helfrich Krebs (?), [mid-June 1702], A I, 21 N. 225, 346.

8   "Tota hac hyeme luctatus sum cum malis quae immiserat autumnus." Leibniz to Hans Sloane, April 17, 1703, British Library, Sloane Ms 4039, pp. 116r–117v; Aiton, "Unpublished Letter," 104; Akademieausgabe reproduction planned for A III, 9.

9   Cf. Johann Barthold Knoche to Leibniz, April 7, 1703, A I 22 N. 33, 48.

10   Leibniz to Queen Sophie Charlotte, March 8, 1703, A I, 22 N. 167, 266.

11   Andrew Fountaine to Leibniz, March 16, 1703, A I, 22 N. 179, 289.

12   Electress Sophie to Leibniz, April 4, 1703, A I, 22 N. 32, 45.

13   Leibniz to Pierre de Falaiseau, April 17, 1703, A I, 22 N. 219, 368–72; Leibniz to Johann Georg Eckhart, April 17, 1703, A I, 22 N. 36, 52–53; Leibniz to Joachim Bouvet, [April 17, 1703], A I, 22 N. 218, 347–68; Leibniz to Jean de Fontaney, [April 17, 1703], A I, 22 N. 220, 372–73. Leibniz's letter of April 17, 1703, to Brosseau has not been found: cf. Christophe Brosseau to Leibniz, April 30, 1703, A I, 22 N. 232, 393–94.

14   Johann Fabricius to Leibniz, April 17, 1703, A I, 22 N. 222, 375–76; Alphonse Des Vignoles to Leibniz, April 17, 1703, A I, 22 N. 221, 374; Carlo Maurizio Vota to Leibniz, April 17, 1703, A I, 22 N. 223, 377–78; Johann Georg Eckhart to Leibniz, April 17, 1703, A I, 22 N. 37, 53–55.

15   For an extensive presentation of the evidence tracing the letter back to this date, see Leibniz to Joachim Bouvet, [April 17, 1703], 347–48. For an alternative argument that dates the letter to May 18, 1703, see Gottfried Wilhelm Leibniz, *Der Briefwechsel mit den Jesuiten in China (1689–1714)*, ed. Rita Widmaier, trans. Malte-Ludolf Babin (Hamburg: Meiner, 2006), N. 40, 397–435 and 732–41.

16   Cf. Wenchao Li and Simona Noreik, eds, *Leibniz und der Gelehrtenhabitus: Anonymität, Pseudonymität, Camouflage* (Cologne: Böhlau Verlag, 2016).

17   Cf. Nora Gädeke, "L'affaire de Monsieur Kortholt oder: Leibniz undercover—Eine Miszelle aus der Praxis der Leibnizedition," *Studia Leibnitiana* 41 (2011): 233–47.

18   "Une taille douce empêche un homme d'aller incognito": Leibniz to Electress Sophie, March 27, 1703, A I, 22 N. 30, 43; Electress Sophie to Leibniz, [late June 1703], A I, 22 N. 45, 68.

19   Leibniz, "Gedancken von Aufrichtung einer Societatis Scientiarum et artium," c. first half of June 1700, A IV, 8 N. 78, 426.

20   Leibniz to Hans Sloane, April 17, 1703.

21   Leibniz to Jakob Heinrich von Flemming, [Berlin], c. April 1703, A I, 22 N. 214, 342. For the crossing out of "Je souhaite" and correction to "M. Cortholt souhaite," see the variant given for line 9. Cf. also Gädeke, "L'affaire de Monsieur Kortholt," 245.

22   "Quant aux affaires de l'Europe, elles sont dans une assiette à nous faire porter envie aux Chinois." Leibniz to Joachim Bouvet, [April 17, 1703].

23   "Bureau d'adresse pour la Chine": Leibniz to Queen Sophie Charlotte, December 14/24, 1697, A I, 14 N. 488, 869; copy of an account by James Cunningham from Chusan to Sloane enclosed in a letter from Hans Sloane

to Leibniz, August 11/22, 1702, LBr. 871, 18–19, 19. (The Sloane letter is slated for inclusion in A III, 9.)

24 Leibniz to Johann Georg Eckhart, April 17, 1703.

25 Leibniz to Joachim Bouvet, [April 17, 1703]; Leibniz to Jean de Fontaney, [April 17, 1703].

26 Leibniz to Joachim Bouvet, [April 17, 1703].

27 For an entry point into the history of binary mathematics, cf. Herbert Breger, "Leibniz' binäres Zahlensystem als Grundlage der Computertechnologie," *Jahrbuch der Akademie der Wissenschaften zu Göttingen* (2008): 385–91. For more on Konrad Zuse and Leibniz, cf. Horst Zuse, "Der lange Weg zum Computer: Von Leibniz' Dyadik zu Zuses Z3," *Vision als Aufgabe: Das Leibniz-Universum im 21. Jahrhundert*, ed. Martin Grötschel et al. (Berlin: Berlin-Brandenburgische Akademie der Wissenschaft, 2016), 111–24.

28 Cf. Ludolf von Mackensen, "Die ersten dekadischen und dualen Rechenmaschinen," in *Gottfried Wilhelm Leibniz. Das Wirken des großen Universalgelehrten als Philosoph, Mathematiker, Physiker, Techniker*, ed. Karl Popp and Edwin Stein (Hanover: Schlütersche, 2000), 85–100.

29 Leibniz, "[Specimen calculi coincidentium et inexistentium]," (1686), LII 4, 7B 2, 60–61; GP 7:236–47.

30 Cf. Theodore Hailperin, "Algebraic Logic: Leibniz and Boole," in *A Boole Anthology: Recent and Classical Studies in the Logic of George Boole*, ed. James Gasser (Dordrecht: Kluwer Academic, 2000), 129–38; and Martin Davis, *The Universal Computer: The Road from Leibniz to Turing* (New York: Norton, 2000), 13–15.

31 Leibniz to Joachim Bouvet, [April 17, 1703], 355.

32 Cf. Mattia Brancato, "Leibniz, Weigel and the Birth of Binary Arithmetic," *Lexicon philosophicum: International Journal for the History of Texts and Ideas* 4 (2016): 151–72.

33 Leibniz to Joachim Bouvet, [April 17, 1703], 354.

34 Leibniz, "Generales inquisitiones de analysi notionum et veritatum," c. spring or late 1686, A VI, 4 A N. 165, 739–88. For more on Leibniz's logical formulas and their significance in the history of logic, cf. Volker Peckhaus, *Logik, Mathesis universalis und allgemeine Wissenschaft: Leibniz und die Wiederentdeckung der formalen Logik im 19. Jahrhundert* (Berlin: Akademie Verlag, 1997).

35 Cf. Robert Kaplan, *The Nothing That Is: A Natural History of Zero* (Oxford: Oxford University Press, 1999); Charles Seife, *Zero: The Biography of a Dangerous Idea* (New York: Viking, 2000); and Sybille Krämer, "Einführung: Wie aus 'nichts' etwas wird: Zur Kreativität der Null," in *Kreativität: XX. Deutscher Kongress für Philosophie 26.–30. September 2005 an der Technischen Universität Berlin*, ed. Günter Abel (Hamburg: Meiner, 2006), 341–43.

36   Leibniz to Duke Rudolph Augustus, January 2/12, 1697, A I, 13 N. 75, 116–21, as well as Leibniz for Rudolph Augustus, "Medaillenentwürfe. Für die Beilage zu N. 75," A I, 13 N. 76, 121–25.

37   Leibniz to Joachim Bouvet, February 15, 1701, A I, 19 N. 202, 401–15.

38   Joachim Bouvet to Leibniz, November 4, 1701, A I, 20 N. 318, 533–55; and Joachim Bouvet for Leibniz. "Darstellung der 64 Hexagramme aus dem Yijing in der Fuxi-Ordnung. Beilage zu N. 318," A I, 20 N. 319, 555–56.

39   Leibniz to Carlo Maurizio Vota, April 4, 1703, A I, 22 N. 203, 321–26.

40   Leibniz to Jean-Paul Bignon, April 7, 1703, A I, 22 N. 207, 332–33.

41   Cf. Rüdiger Glaser, *Klimageschichte Mitteleuropas: 1200 Jahre Wetter, Klima, Katastrophen*, 2nd ed. (Darmstadt: Wissenschaftliche Buchgemeinschaft, 2008), 176.

42   Leibniz to Joachim Bouvet, April 17, 1703, 360.

43   "Renaissance-Magus" is the title given to him by Peter Sloterdijk. Peter Sloterdijk, *Philosophische Temperamente: Von Platon bis Foucault* (Munich: Diederichs, 2009), 59.

44   Carlo Maurizio Vota to Leibniz, April 17, 1703, 377.

45   See commentary on Leibniz's letter to Hans Sloane, April 17, 1703, planned for A III, 9.

46   Cf. Leibniz to Conrad Barthold Behrens, [Febuary 6, 1696], A I, 12 N. 264, 397. There is no record of what foods Leibniz ate on Brüderstraße. A comparison can be made to his food consumption at his lodgings in Wolfenbüttel. From there, there is at least a receipted bill, though the items listed are rather meager: cf. Johann Christoph Balcke for Leibniz, "Quittierte Rechnung," February 14/24, 1699, A I, 16, N. 66, 104–5. I thank Malte-Ludolf Babin for the tip.

47   Leibniz to Joachim Bouvet, [April 17, 1703], 364; Leibniz to Johann Georg Eckhart, April 17, 1703, 52–53.

48   Leibniz to Joachim Bouvet, [April 17, 1703], 358.

49   Jost rightly stresses that the significance of his thinking in structural correspondences and analogs has received too little attention in Leibniz scholarship: Jürgen Jost, *Leibniz und die moderne Naturwissenschaft* (Berlin: Springer, 2019), 6–9. Cf. also Constanze Peres, "Neuheit und Kreativität—analogisches Denken und Leibniz' Idee der Erfindung," in *Wie entsteht Neues? Analogisches Denken, Kreativät und Leibniz' Idee der Erfindung*, ed. Constanze Peres (Paderborn: Wilhelm Fink Verlag, 2020), 7–12.

50   Cf. Leibniz, "Dissertatio exoterica de usu geometriae, et statu praesenti, ac novissimis ejus incrementis, [August–September 1676], A VII, 6 N. 49_1, 483–514; as well as Leibniz, "De arte characteristica inventoriaque analytica combinatoriave in mathesi universali," c. May 1679–April 1680, A VI, 4. Teil A N. 78, 315–32. Leibniz gives an example of the latter quality, his

combinatorial ingenuity, by pointing to his having expanded on Huygens's
pendulum clock design and conceived of portable clocks simply by applying
the theoretical principles of mechanics to the realm of physical technology.

51 Leibniz to Hans Sloane, April 17, 1703.

52 Cf. the explanatory notes by Widmaier and Babin in Leibniz, *Der Briefwechsel mit den Jesuiten in China*, 740.

53 Leibniz to Joachim Bouvet, [April 17, 1703], 358.

54 Kurt Gödel, "Über formal unentscheidbare Sätze der Principia Mathematica und verwandter Systeme I," *Monatshefte für Mathematik und Physik* 38
(1931): 173–98.

55 Leibniz to Joachim Bouvet, [April 17, 1703], 359.

56 Cf. Rita Widmaier, "Die Dyadik in Leibniz' letztem Brief an Nicolas
Remond," *Studia Leibnitiana* 49, no. 2 (2017): 139–76.

57 This is according to the eyewitness account of Johann Hermann Vogler,
Leibniz's last servant and amanuensis: Paul Ritter, "Bericht eines Augen-
zeugen über Leibnizens Tod und Begräbnis," *Zeitschrift des Historischen
Vereins für Niedersachsen* 81 (1916): 251.

58 Leibniz to Joachim Bouvet, [April 17, 1703], 356; Ludwig Wittgenstein,
Philosophische Untersuchungen, vol. 2 of *Werkausgabe* (Frankfurt: Suhrkamp,
1984), 378.

Chapter 5: HANOVER, JANUARY 19, 1710

1 Leibniz, "La place d'autruy," c. 1679, A IV, 3 N. 137, 903: "La place d'autruy
est le vray point de perspective en politique aussi bien qu'en morale."

2 Zacharias Konrad von Uffenbach, *Merkwürdige Reisen durch Niedersachsen, Holland und England* (Ulm/Memmingen, 1753), 1:410–11, 418–21,
and 437–42.

3 Cf. Markus Friedrich and Monika Müller, eds., *Zacharias Konrad von Uffenbach: Büchersammler und Polyhistor in der Gelehrtenkultur um 1700* (Berlin:
Walter de Gruyter, 2020).

4 Uffenbach, *Merkwürdige Reisen*, 1:409–10.

5 Uffenbach, *Merkwürdige Reisen*, 1:418–21.

6 Uffenbach, *Merkwürdige Reisen*, 1:421–37.

7 Uffenbach, *Merkwürdige Reisen*, 1:409.

8 For these and further examples, cf. Ekkehard Göhrlich, "Leibniz als Mensch
und Kranker," medical diss. (Hanover, 1987), 223.

9 Cf. Kurt Müller and Gisela Krönert, eds., *Leben und Werk von Gottfried
Wilhelm Leibniz: Eine Chronik* (Frankfurt: Klostermann, 1969), 204, which
cites a February 19, 1707, letter from Christian Thomasius to Leibniz.

10   Leibniz, "Grundriss eines Bedenkens von Aufrichtung einer Societät," c.
     1671, A IV, 1 N. 43, 539.

11   Leibniz, "Entwurff gewisser Staats-Tafeln," [Spring 1680], A IV, 3 N.
     29, 240–349.

12   Uffenbach, *Merkwürdige Reisen*, 1:415.

13   For more on the audience chamber in Leibniz's house, cf. Cord Meckseper,
     Das Leibnizhaus in Hannover: Die Geschichte eines Denkmals (Hanover:
     Schlütersche, 1983), 63–64.

14   Cf. Nora Gädeke, ed., *Leibniz als Sammler und Herausgeber historischer Quel-
     len* (Wiesbaden: Harrasowitz Verlag, 2012).

15   Leibniz, "[Über die Exemtion adliger Güter]," in *Gottfried Wilhelm Leibniz:
     Schriften und Briefe zur Geschichte*, trans. Malte-Ludolf Babin and ed. Gerd
     van den Heuvel (Hanover: Hahnsche, 2004), N. 27, 515–17. (Enclosed in a
     letter from Leibniz to Andreas Gottlieb von Bernstorff, January 19, 1710.)

16   Cf. Horst Eckert, *Gottfired Wilhelm Leibniz' Scriptores Rerum Burns-
     vicensium: Entstehung und historiographische Bedeutung* (Frankfurt: V.
     Klostermann, 1971).

17   Uffenbach, *Merkwürdige Reisen*, 1:438.

18   Leibniz, *Nouveaux essais sur l'entendement humain*, A VI, 6 N. 2, 367–72 (bk.
     IV, chap. 2, §5–§13).

19   Gerd van den Heuvel, "Einleitung," in *Gottfried Wilhelm Leibniz: Schriften
     und Briefe zur Geschichte*, trans. Malte-Ludolf Babin and ed. Gerd van den
     Heuvel (Hanover: Hansche, 2004), 35.

20   Uffenbach, *Merkwürdige Reisen*, 1:438.

21   Bartholomäus Des Bosses to Leibniz, January 18, 1710. Leibniz, *Der Brief-
     wechsel mit Bartholomäus Des Bosses*, trans. and ed. Cornelius Zeheter (Ham-
     burg: F. Meiner, 2007), 164–71; Uffenbach, *Merkwürdige Reisen*, 1:438.

22   For further commentary on the transmission of the story through Marti-
     nus Polonus, cf. Elisabeth Gössmann, *"Die Päpstin Johanna": Der Skandal
     eines weiblichen Papstes: Eine Rezeptionsgeschichte*, 3rd ed. (Munich: Aufbau
     Taschenbuch Verlag, 1998), 33–42.

23   Leibniz, "Flores sparsi in tumulum papissae," in *Bibliotheca Histor-
     ica Goettingensis*, ed. Christian Ludwig Scheidt (Göttingen/Hanover,
     1758), 1:297–392.

24   Eckhart, "Lebensbeschreibung," in *Leibniz-Biografien*, ed. J. A. Eberhard
     and J. G. Eckhart (Hildesheim: George Olms, 1982), 201; Christoph Got-
     tlieb von Murr, "Einige Zusätze zum Eckhartischen Lebenslaufe des Herrn
     von Leibnitz" (1779), in Eberhard and Eckhart, *Leibniz-Biografien*, 219.

25   Johann Jacob Julius Chuno to Leibniz, January 7, 1710, LBr. 185, Bl. 103–
     4; Johann Theodor Jablonski to Leibniz, December 21, 1709, LBr. 440 Bl.
     130–31, available online: T 1709.

26   Murr, "Einige Zusätze," 211.

27   How hard the news of her death hit him is evident above all in his letters, collected in A I, 24. Long into the summer of 1705, Leibniz, for sheer grief, is barely capable of continuing his correspondence; cf. for example Leibniz to William Wotton, July 10, 1705, A I, 24 N. 434, 777; and Leibniz to Antonio Magliabechi, July 21, 1705, A I, 24 N. 446, 802.

28   This is clear from Troyel's letters from this period. Isaac Troyel to Leibniz, November 23, 1709, LBr. 942 Bl. 4; Isaac Troyel to Leibniz, January 11, 1710, LBr. 942 Bl. 5; Isaac Troyel to Leibniz, January 21, 1710, LBr. 942 Bl. 6.

29   Leibniz, *Essais de Théodicée sur la bonté de dieu, la liberté, de l'homme et l'origine du mal* (Amsterdam, 1710); also GP 6. Citations in this passage are taken from the French-German edition: Leibniz, *Die Theodizee von der Güte Gottes, der Freiheit der Menschen und dem Ursprung des Übels*, vols. 2/1 and 2/2 of *Philosophische Schriften*, trans. and ed. Herbert Herring (Frankfurt: Suhrkamp, 1996). For an English translation, see Leibniz, *Theodicy, Abridged*, trans. E. M. Huggard (Ontario: J.M. Dent & Sons, 1966).

30   Immanuel Kant, "Über das Mißlingen aller philosophischen Versuche in der Theodizee," in *Schriften zur Anthropologie, Geschichtsphilosophie, Politik und Pädagogik*, vol. 9, pt. 1 of *Werke*, ed. Wilhelm Weischedel (Darmstadt: Wissenschaftliche Buchgesellschaft, 1983), 105–24.

31   Leibniz, *Theodizee*, vol. 2/1, 50–51; and 3. Teil, §405–§417, vol. 2/2, 247–69.

32   Lorenzo Valla, *Über den freien Willen. De libero arbitrio*. ed. Eckhard Keßler (Munich: W. Fink, 1987), 103–19 (pp. 31–39 in the Latin edition of 1934). The oldest manuscript dates back to 1461.

33   Cf. Leibniz to Electress Sophie, March 4, 1707 and enclosed document, in Leibniz and Electress Sophie of Hanover, *Briefwechsel*, ed. Wenchao Li, trans. Gerda Utermöhlen and Sabine Sellschopp (Göttingen: Wallstein Verlag, 2017), N. 309, 310, 626–28.

34   Leibniz, "Chansons et autres vers pour le festin de trimalcion moderne," c. February 25, 1702, A IV, 9 N. 113, 788–900. Cf. also Malte-Ludolf Babin, "Leibniz und der trimalcion moderne: Edition der Berichte von der Aufführung im Februar 1702," in *Studien zu Petron und seiner Rezeption / Studie su Petronio e sulla sua fortuna*, ed. Luigi Castagna and Eckhard Lefèvre (Berlin: Walter de Gruyter, 2007), 331–60.

35   Uffenbach, *Merkwürdige Reisen*, 1:411–14.

36   Leibniz to Bartholomäus des Bosses, April 24, 1709, in Leibniz and Des Bosses, *Der Briefwechsel*, trans. and ed. Zeheter, N. 52, 126–32; Bartholomäus Des Bosses to Leibniz, January 18, 1710, *Der Briefwechsel*, N. 65, 164–71; Leibniz to Bartholomäus Des Bosses, January 1710, *Der Briefwechsel*, N. 66, 171–74.

37   Cf. Wilhelm Herse, "Leibniz und die Päpstin Johanna," in *Beiträge zur Leibniz-Forschung*, ed. Georgi Schischkoff (Reutlingen: Gryphius-Verlag, 1947), 153.

38   Cf. Peter Fenves, *Arresting Language: From Leibniz to Benjamin* (Stanford: Stanford University Press, 2001), 76–77.

39   Leibniz, "Flores sparsi in tumulum papissae," 1:342.

40   Cf. Leibniz, "Stoffe zu historischen Romanen," in Babin and Van den Heuvel, *Schriften und Briefen zur Geschichte*, N. 35, 581–82.

41   Leibniz, "Über die Kontingenz," in *Kleine Schriften zur Metaphysik*, trans. and ed. Hans Heinz Holz (Frankfurt: Suhrkamp, 1996), 185. In the Academy edition: "De Contingentia," A VI, 4. Teil B N. 325, 1651. See also Leibniz, "De libertate, contingentia et serie causarum, providentia," A VI, 4. Teil B N. 326, 1653–54.

42   Cf. Gössmann, *Die Päpstin Johanna*, 258–59.

43   Cf. Willy Kabitz, *Leibniz und Berkeley* (Berlin: Sitzungsberichte der Preußischen Akademie der Wissenschaften, Phil. Hist. Klasse, 1932).

44   Leibniz, *Theodizee*, pt. 3, § 416, vol. 2/2, 266: "Le crime de Sextus sert à de grandes; il rend Rome libre, il en naître un grand empire qui donnera de grands exemples." *Trans.*: The passage in quotes is taken from Leibniz, *Theodicy, Abridged*, 175.

45   Leibniz, "Flores sparsi in tumulum papissae," 1:364.

46   Fenves, *Arresting Language*, 74.

47   Uffenbach, *Merkwürdige Reisen*, 1:442; regarding the "sturdy repast": Eckhart, "Lebensbeschreibung," 197.

48   Konrad Franke, "Zacharias Conrad von Uffenbach als Handschriftensammler. Ein Beitrag zur Kulturgeschichte des 18. Jahrhunderts," *Börsenblatt für den deutschen Buchhandel, Frankfurter Ausgabe* 51 (June 29, 1965): 1249.

Chapter 6: VIENNA, AUGUST 26, 1714

1   Leibniz to Vincent Placcius, February 21 / March 2, 1696, A II, 3 N. 48, 139: "Scilicet qui me non nisi editis novit, non novit."

2   Cf. Margot Faak, *Leibniz als Reichshofrat* (Berlin: Springer Spektrum, 2016).

3   "Commercium epistolicum D. Johannis Collins, et aliorum de analysi promota: jussu societatis regiae in lucem editum," London 1712/13. At the Gottfried Wilhelm Leibniz Bibliothek, there is a copy with numerous notes in the margins made by Leibniz: GWLB, MS IV 379a.

4   Leibniz to Charlotte Elisabeth von Klencke, October 4, 1713, LK-MOW Klencke 10 Bl. C 74–75 (formerly: LH 51 9 Bl. 74–75) (draft), in T 1713.

5 For background on the plague in Vienna, cf. Heide Schmölzer, *Die Pest in Wien* (Innsbruck: Haymon Verlag, 2015), 197–204.

6 Cf. Leibniz, "Puncta so eine schleünige anstalt bedürffen," c. 1713–14, LH 35, Bl. 119–20 (fair copy in Leibniz's hand); Bl. 121 (draft). For additional context, see also Michael Kempe, "Dr. Leibniz, oder wie ich lernte, die Bombe zu lieben. Zum Verhältnis von Wissenschaft und Militärtechnik in Europa um 1700," in *Der Philosoph im U-Boot: Praktische Wissenschaft und Technik im Kontext von Gottfried Wilhelm Leibniz*, ed. Michel Kempe (Hanover: Gottfried Wilhelm Leibniz Bibliothek, 2015), 113–45.

7 Gottfried Wilhelm Leibniz for Emperor Charles VI, "[Promemoria betreffend Baracken für ansteckende Kranke]," LH 34 Bl. 127. Cf. also Georg Fischer, "Der Philosoph Leibniz über Baracken," *Deutsche Zeitschrift für Chirurgie* 19 (1884): 135–36.

8 For more on this development, cf. Kai Kauffmann, *"Es ist nur ein Wien!" Stadtbeschreibungen von Wien 1700 bis 1873. Geschichte eines literarischen Genres der Wiener Publizistik* (Vienna: Böhlau, 1994). I thank Thomas Stockinger for the tip.

9 Cf. Johann Friedrich Hodann to Leibniz, July 22, 1714, LBr. 411 Bl. 525–26; also accessible in T 1714.

10 Cf. Richard Doebner, ed., *Leibnizens Briefwechsel mit dem Minister von Bernstorff und andere Leibniz betreffende Briefe und Aktenstücke aus den Jahren 1705–1706* (Hanover: Hahn, 1882), 13.

11 Johann Georg Eckhart to Leibniz, June 24, 1714, in Doebner, *Leibnizens Briefwechsel*, 82–83.

12 Johann Friedrich Hodann to Leibniz, July 22, 1714, LBr. 411 Bl. 525–26; T 1714.

13 Leibniz to Matthias Johann von der Schulenburg, August 25, 1714, Staatsbibliothek zu Berlin, Preußischer Kulturbesitz, Ms. Savigny 38 Bl. 112–13; accessible in T 1714.

14 Johann Friedrich Hodann to Leibniz, August 16, 1714, LBr. 411 Bl. 537; transcription available in T 1714; Leibniz to Johann Friedrich Hodann, August 25 1714, Kongelige Bibliotek Kopenhagen, NKS 4° Nr. 104; transcription available in T 1714.

15 Johann Friedrich Hodann to Leibniz, August 19, 1714, LBr. 411 VBl. 538–39; transcription available in T 1714.

16 Leibniz to Andreas Gottlieb von Bernstorff, August 22, 1714, in Doebner, *Leibnizens Briefwechsel*, 89–90; Leibniz to Andreas Gottlieb von Bernstorff, August 24, 1714, *Leibnizens Briefwechsel*, 90–91; Leibniz to Matthias Johann von der Schulenburg, August 25, 1714, Staatsbibliothek zu Berlin, Preußischer Kulturbesitz, Ms. Savigny 38 Bl. 112–13; transcription available

in T 1714. One learns from the postscript that it is too late to send the letter off on August 22 (Wednesday).

17   Leibniz to Johann Friedrich Hodann, August 25, 1714, Kongelige Bibliotek Kopenhagen, NKS 2753 4° Nr. 104; transcription available in T 1714.

18   Johann Georg Eckhart to Leibniz, August 19, 1714, LBr. 311 Bl. 24–25.

19   Elector George Louis for Johann Georg Eckhart, August 25, 1714, Staatsarchiv Wolfenbüttel, 5633 Bl. 99–102.

20   John Ker of Kersland to Leibniz, August 25, 1714, in John Ker of Kersland, *The Memoirs of John Ker, of Kersland . . .* (London, 1726–27), 94–95.

21   Leibniz to Matthias von der Schulenburg, August 25 1714. It is the postscript referred to here. The bound anthology is in Vienna at the Österreichische Nationalbibliothek Wien, Cod. Pal. Vindob. 10588.

22   *Wienerisches Diarium*, August 25–28, 1714, Nr. 1155.

23   Leibniz to John Chamberlayne, August 25, 1714, in Pierre Des Maizeaux, ed., *Recueil de diverses pieces, sur la philosophie, la religion naturelle, l'histoire, les mathematiques, &c.* (Amsterdam, 1720), 2:123–24.

24   Charles Hugony to Leibniz, April 17, 1714, GWLB, LBr 430, Bl. 23–24.

25   For more on the history of the composition of the *Monadologie* and the *Principes* in their respective manuscript versions, cf. Clara Strack, *Ursprung und sachliches Verhältnis von Leibnizens sogenannter Monadologie und den Principes de la nature et de la grâce. 1. Teil: Die Entstehungsgeschichte der beiden Abhandlungen* (Berlin, 1915). For more on the collation of the manuscripts of both treatises see Leibniz, *Sogenannte Monadologie und Principes de la nature et de la grâce*, ed. Clara Strack (Berlin: Walter de Gruyter, 1967). Cf. also Antonio Lamarra, Roberto Palaia, and Pietro Pimpinella, *Le Prime traduzioni della Monadologie di Leibniz (1720–1721): Introduzione storico-critica, sinossi dei testi, concordanze contrastive* (Florence: Leo S. Olschki, 2001).

26   Leibniz to Nicolas-François Rémond, August 26, 1714, GP 3:624. Cf. extract in Leibniz's hand: LBr. 768, Bl. 22.

27   For more on the taverns and markets in Vienna's city center, see Antonio Bormastino, *Historische Erzehlung von der Käyserlichen Residentz-Stadt Wienn . . .* (Vienna, 1715), 144–50.

28   Patrick Riley, *Leibniz' "Monadologie" als Theorie der Gerechtigkeit* (Hanover: Wehrhahn Verlag, 2017), makes reference to the fair copy of the lecture in the handwriting of a copyist, LH 39 Bl. 58–61r.

29   Leibniz, "Spatium absolutum . . . ," [August–September 1714], LH 35, 1, 12 Bl. 6r and 19r; published in the appendix of Vincenzo De Risi, *Geometry and Monadology: Leibniz's "Analysis Situs" and Philosophy of Space* (Basel: Birkhäuser, 2007, N. 11, 608–11. *Trans.:* In translating this passage, I have also drawn from Vincenzo De Risi, "Analysis Situs, the Foundations of Mathematics, and a Geometry of Space," in *The Oxford Handbook*

of Leibniz, ed. Maria Rosa Antognazza (Oxford: Oxford University Press, 2013), 247–58.

30   Cf. Vincenzo De Risi, *Leibniz on the Parallel Postulate and the Foundations of Geometry: The Unpublished Manuscripts* (Birkhäuser: Springer, 2016).

31   Gottfried Teuber to Leibniz, July 30, 1714, LBr. 916 Bl. 16; transcription available in T 1714.

32   Leibniz, "Extrait du dictionnaire de M. Bayle article Rorarius p. 2599 sqq. De l'edition de l'an 1702 avec mes remarques," GP 4:524–54. Gerhardt dated the text back to 1702. In fact, however, Leibniz completed it in the first half of 1714 in Vienna and included it in the prince's anthology: Leibniz, "Objections de M. Bayle avec les résponses de l'auteur du système," Österreichische Nationalbibliothek Wien, Cod. 10588, Bl. 128–206. For more on the dating, cf. Strack, *Ursprung,* 31–33. For more on Leibniz as an early thinker on autonomous driving, see the illuminating passage in Stephan Meder, *Der unbekannte Leibniz: Die Entdeckung von Recht und Politik durch Philosophie* (Vienna: Böhlau Verlag, 2018), 184–85.

33   Gottfried Wilhelm Leibniz, "Facilius intelligere deum, comprehendere corpus," c. mid-1685 to late 1688, A VI, 4, Part A, N. 150, 610.

34   "Le present est gros de l'avenir, le futur se pouvoit lire dans le passé, l'eloigné est exprimé dans le prochain. On pourroit connoitre la beauté de l'univers dans chaque ame, si l'on pouvoit deplier tous ses replis, qui ne se developpent sensiblement qu'avec le temps." Leibniz, "Principes de la nature et de la grâce fondés en raison," §13, GP 6:598–606. For an English translation of the *Principles*, see Leibniz, *Philosophical Writings*, trans. Mary Morris and G.H.R. Parkinson, 195–204.

35   Leibniz, untitled ms. containing the so-called "Monadologie," GP 6:607–23, §7: "Les Monades n'ont point de fenêtres, par lesquelles quelque chose y puisse entrer ou sortir." For an English translation, see Leibniz, *Leibniz's Monadology*, trans. Lloyd Strickland (Edinburgh: Edinburgh University Press, 2014).

36   Cf. Jürgen Jost, *Leibniz und die moderne Naturwissenschaft* (Berlin: Springer, 2019), 157–58.

37   René Descartes, *Meditations de prima philosophia / Meditationen über die erste Philosophie*, trans. Gerhart Schmidt (Stuttgart: Reclam, 1986), "Zweite Meditation," §13, 92–93.

38   Pierre Bayle, "Pensées diverses I, §V," *Œuvres diverses, tome troisième, première partie* (Hague: 1727), 11 B. My thanks to Stephan Lorenz for this tip.

39   For more on the wet and humid summer of 1714 in Austria and Vienna, cf. Elisabeth Strömmer, *Klima-Geschichte: Methoden der Rekonstruktion und historische Perspektive: Ostösterreich 1700 bis 1830* (Vienna: Deuticke, 2003), 109–10. My thanks to Thomas Stockinger for the tip.

40   Leibniz to Emperor Charles VI, August 24, 1714, LH 13 Bl. 155–56 (draft).

41   For more on the figure of the double agent, cf. Marika Keblusek and Vera
     Noldus Badeloch, eds., *Double Agents: Cultural and Political Brokerage in
     Early Modern Europe* (Leiden: Brill, 2011).

42   Johann Friedrich Hodann to Leibniz, August 19, 1714, LBr. 411 Bl. 538–39;
     transcription available in T 1714.

43   Leibniz to Johann Friedrich Hodann, August 29, 1714, Kongelige Bibliotek
     Kopenhagen, NKS 2753 4° Nr. 105; transcription available in T 1714.

Chapter 7: HANOVER, JULY 2, 1716

1   Leibniz to Nicolas-François Rémond, February 11, 1716, GP 3:634–40: "Ce
    qui est un bonheur dans le malheur."

2   Cf. Nora Gädeke, "Leibniz' Korrespondenz im letzten Lebensjahr: Gerber
    Reconsidered," in *1716—Leibniz' letztes Lebensjahr: Unbekanntes zu einem
    bekannten Universalgelehrten*, ed. Michael Kempe (Hanover: Gottfried Wil-
    helm Leibniz Bibliothek, 2016), 83–109.

3   Leibniz to Theobald Schöttel, July 2, 1716, Österreichische National-
    bibliothek Wien, ser. Nov. 11992 Nr. 27, transcription in T 1716; Leibniz
    to Joseph Schöttel (?), [July 2, 1716], Österreichische Nationalbibliothek
    Wien, ser. Nov. 11992 Nr. 28, transcription in T 1716; Leibniz to Ferdidand
    Orban S.J., July 2, 1716, Staatsbibliothek Berlin, Ms. Lat. 311 D; Leibniz
    to Louis Bourguet, July 2, 1716, Universitaire Bibliotheken Leiden, Ms.
    293 B Bl. 273–74; Bibliothèque municipal Rouen, O 39 pp. 331r–331V
    (copy), GP 3:594–96; Leibniz to Rinaldo d'Este, Duke of Modena, July
    2, 1716, LBr. 676 Bl. 503 (retained copy); Archivio di Stato di Modena,
    Letterati B 31. Fasz. Leibniz. Pp. 5–6, transcription in T 1716; Leibniz to
    Ludovico Antonio Muratori, July 2, 1716, LBr. 676 Bl. 504 (retained copy);
    Biblioteca Estense Modena, Archivio Muratoriano Filza 85, fasc. 6 (received
    letter); LBr. 676 Bl. 504a–504b (copy of retained copy), transcription in
    T 1716.

4   For more on Leibniz's experiments with magic squares and cubes in 1715–
    16, cf. Christian Boyer, "Le plus petit cube magique," *La recherche: L'actualité
    des sciences* 373 (March 2004): 48–50.

5   "Je ne saurois assés admirer la vivacité et le jugement de ce grand Prince. Il
    fait venir des habiles gens de tous côtés, et quand il leur parle, ils en sont
    tout etonnés, tant il leur parle à propos." Leibniz to Louis Bourguet, July 2,
    1716, GP 3:596.

6   Cf. Gisela Piel and Susanne Luber, "Ein 'berühmtes Brunnenjahr.' Zar
    Peter I. in Pyrmont im Juni 1716," in *Zar Peter der Grosse: Die zweite große*

Reise nach Westeuropa, 1716–1717: Auf Europäischer Bühne, ed. Dieter Alfter (Hameln: Niemeyer, 1999), 87–96.

7   Leibniz to Louis Bourguet, July 2, 1716, GP 3:596: "Il s'informe de tous les arts mecaniques: mais sa grand curiosité est pour tout ce qui a du rapport à la navigation; et par consequent it aime aussi l'Astronomie et la Geographie. J'espere que nous apprendrons par son moyen, si l'Asie est attachée à l'Amérique." For more on Leibniz's Russia plans, cf. Regina Stuber, "Leibniz' Bemühungen um Russland: eine Annäherung," in Kempe, *1716—Leibniz' letztes Lebensjahr*, 203–39.

8   Leibniz, "Concept einer Denkschrift Leibniz's über die Verbesserung der Künste und Wissenschaften im Russischen Reich," in Woldemar Guerrier, *Leibniz in seinen Beziehungen zu Russland und Peter dem Grossen: Eine geschichtliche Darstellung dieses Verhältnisses nebst den darauf bezüglichen Briefen und Denkschriften* (St. Petersburg, 1873), 351–52. Manuscript at the GWLB, MS 33, 1749, Bl. 136–43.

9   For more on Leibniz's information gathering on Russia and China, cf. Christine Roll, "Barbaren? 'Tabula rasa?' Wie Leibniz sein neues Wissen über Russland auf den Begriff brachte; eine Studie über die Bedeutung und Vernetzung gelehrter Korrespondenzen für die Ermöglichung aufgeklärter Diskurse," in *Umwelt und Weltgestaltung: Leibniz' politisches Denken in seiner Zeit*, ed. Friedrich Beiderbeck et al. (Göttingen: Vandenhoek & Ruprecht, 2014), 307–58; and Michael C. Carhart, *Leibniz Discovers Asia: Social Networking in the Republic of Letters* (Baltimore: John Hopkins University Press, 2019).

10 Nicolaas Witsen, *Nieuwe Lantkaarte van het noorder en ooster dell van Asa en Europa, strekkende van Nova Zemla tot China* (Amsterdam, 1687), GWLB, Ktr. 118. The GWLB map is presumably the copy that Leibniz acquired in Holland and later used; cf. Leibniz to Nicolaas Witsen, March 26, 1694, A I, 10 N. 211, 338–40.

11 Cf. Michael Kempe, " 'Schon befand ich mich in Gedanken in Rußland . . .': Johann Jakob Scheuchzer im Briefwechsel mit Gottfried Wilhelm Leibniz," in *Alte Löcher—neue Blicke. Zürich im 18. Jahrhundert: Aussen- und Innenperspektiven*, ed. Helmuth Holzhey and Simone Zurbuchen (Zurich: Chronos-Verlag, 1997), 283–97.

12 Leibniz, "Denkschrift über die Collegien," *Leibniz in seinen Beziehungen zu Russland und Peter dem Grossen*, 364.

13 For more on this cf. Stephan Meier-Oeser, " 'Adieu le Vuide, les Atomes, et toute la Philosophie de M. Newton': Zur Leibniz-Clarke-Kontroverse," in Kempe, *1716—Leibniz' letztes Lebensjahr*, 293–317; and Monike Meier, "Leibniz' Briefwechsel mit Caroline von Ansbach, Princess of Wales— 'Leibniz' in Großbritannien und neue ökumenische Initiativen," in *1716—Leibniz' letztes Lebensjahr*, 241–91.

14   Cf. Meier-Oeser, "Adieu le Vuide," 317.

15   Leibniz to Bourguet, July 2, 1716, GP 3:595; cf. the German translation:
Volkmar Schüller, ed., *Der Leibniz-Clarke Briefwechsel* (Berlin: Akademie
Verlag, 1991), 256ff.

16   Cf. Ursula Goldenbaum, "Eine frühe Rezeption von Leibniz' Dynamik
oder was wir von der Korrespondenz zwischen Louis Bourguet und Jacob
Hermann lernen können," in *Leibniz in Latenz: Überlieferungsbildung als
Rezeption (1716–1740)*, ed. Nora Gädeke and Wenchao Li (Stuttgart: Franz
Steiner Verlag, 2017), 223–53.

17   Cf., for instance, Louis Bourguet to Leibniz, March 31, 1716, LBr. 103, Bl.
44–45; transcription in T 1716; Leibniz to Louis Bourguet, April 20, 1716,
GP 3:594; Louis Bourguet to Leibniz, May 15, 1716, in Niedersächsische
Staats- und Universitätsbibliothek Göttingen, 2 Cod. Ms. Philos. 138 Bl.
15–16; transcription in T 1716.

18   Cf. Rüdiger Glaser, *Klimageschichte Mitteleuropas: 1200 Jahre Wetter, Klima,
Katastrophen*, 2nd ed. (Darmstadt: Wissenschaftliche Buchgemeinschaft,
2008), 179.

19   Leibniz to Louis Bourguet, July 2, 1716, GP 3:595.

20   Louis Bourguet to Leibniz, April 7, 1714, GWLB, MS 23, 23a Bl. 90–95;
Leibniz to Louis Bourguet, [May 1714], Österreichische Nationalbibliothek
Wien, Cod. Pal. Vindob. 10588, Bl. 108–27; transcription in T 1714. For
more on the two letters, cf. Federico Silvestri, "Due inediti dal carteggio
Leibniz-Bourguet," *Rivista di storia della filosofia, nuova seria 3* (2018): 481–
505. Further textual witnesses to May letter from Leibniz to Bourguet: LBr.
817 Bl. 64–65 (draft by Leibniz), Ms 23, 23a Bl. 96–103 (fair copy in a
scribe's hand with handwritten changes in Leibniz's hand), LBr. 817 Bl. 66–
71 (fair copy in a scribe's hand with handwritten changes in Leibniz's hand).

21   Kurt Müller and Gisela Krönert, eds., *Leben und Werk von Gottfried Wilhelm
Leibniz: Eine Chronik* (Frankfurt: Klostermann, 1969) contains an account
by a certain Hofrat Schläger, looking back on his childhood (250). For more
on the carriage and Leibniz's appearance, cf. 206.

22   Leibniz, "[Sur la generation des insectes et autres petits animaux]," (1714),
LH 37, 7 Bl. 6–7. "Mais on s'est apperçu aujourdhuy que certaines grandes
mouches ou guespes, asses ressemblantes à des mouches à miel persecutent
les boeufs, et leur perçant le cuir y deposent des oeufs, d'où vient un veer,
qui se nourrissant sous la peau, se change enfin en mouche semblable à celle
dont il etoit né." Cf. the edition of this text with commentary: Osvaldo
Ottaviani and Alessanro Becchi, eds., "Leibniz on Animal Generation: An
Unpublished Text (LH 37, 7 Bl. 6–7) with Introduction, Critical Edition,
Translation, and Commentary," *Leibniz Review* 30 (2020): 63–106.

23   Cf. James G. O'Hara, " 'J'aime mieux un Leewenhoek qui me dit ce qu'il

voit, qu'un Cartesien qui me dit ce qu'il pense,' Leibniz, Leeuwenhoek und
die Engwicklung der Experimentellen Naturwissenschaft," in Kempe,
1716—Leibniz' letztes Lebensjahr, 145–75.

24   Cf. Daniel Garber, "Steno, Leibniz, and the History of the World," in *Steno and the Philosophers,* ed. Raphaële Andrault and Mogens Lærke (Leiden: Brill, 2018), 201–29.

25   Leibniz, "Protogaea autore G.G.L.," *Acta eruditorum* (January 1693): 40–42; Louis Bourguet to Leibniz, April 7, 1714; Leibniz to Louis Bourguet, [May 1714]: "Et de Salamandres (pour ainsi dire) seroient devenus des poissons, et puis des amphibies, et enfin des animaux terrestres et volatiles."

26   Leibniz's "Primordial Earth" wasn't published until 1749: Leibniz, *Protogaea. Sive de prima facie telluris et antiquissimae historiae vestigiis in ipsis naturae monumentis dissertatio . . .* (Göttingen, 1749).

27   "La mort peut ramener aux salamandres et peut être à une subtilité au delà." Leibniz to Louis Bourguet, [May 1714].

28   Cf. Richard T. W. Arthur, "Leibniz, Organic Matter and Astrobiology," in *Tercentenary Essays on the Philosophy and Science of Leibniz,* ed. Lloyd Strickland et al. (Cham: Palgrave Macmillan, 2017), 81–107. For more on a general theory of evolution, cf. also Addy Pross, "Toward a General Theory of Evolution: Extending Darwinian Theory to Inanimate Matter," *Journal of Systems Chemistry* 2, no. 1 (2011): 1–14.

29   Leibniz to Bourguet, August 5, 1715, GP 3:582.

30   Cf. Gerd van den Heuvel, "Die germanische Frühzeit als beste der möglichen Welten? Historiografie und Geschichstsauffassung bei Justus Möser und Gottfried Wilhelm Leibniz," in *Perspektiven der Landesgeschichte: Festschrift für Thomas Vogtherr,* ed. Christine van den Heuvel et al. (Göttingen: Wallstein Verlag, 2020), 409–26.

31   Cf. Louis Bourguet to Leibniz, October 6, 1715, LBr. 103, Bl. 34–35, GP 3:583–86; Louis Bourguet to Leibniz, February 7, 1716, LBr. 103 Bl. 38–41, transcription in T 1716; Leibniz to Louis Bourguet, February 24, 1716, Bibliothèque municipal Rouen, O 39 pp. 295r–298v, GP 3:591; Louis Bourguet to Leibniz, March 16, 1716, LBr. 103 Bl. 47–50, T 1716; and Leibniz to Louis Bourguet, April 3, 1716, Universitaire Bibliotheken Leiden, Ms. 293 B Bl. 269–70, GP 3:591–93.

32   Leibniz to Rudolf Christian von Bodenhausen, December 20/30, 1693, A III, 5 N. 201, 672. For more on the pouncing tiger image, see also the epilogue in James O'Hara, *Leibniz's Correspondence in Science, Technology and Medicine (1676–1701)* (Leiden: Brill, 2024).

33   Leibniz to Louis Bourguet, July 2, 1716, GP 3:595. Cf. also Leibniz to Louis Bourguet, April 3, 1716, GP 3:591–93.

34   Leibniz to Johann Fabricius, July 6, 1716, first printing based on the sent

letter, which has not been found: Christian Kortholt, *Epistolae ad diversos* ... (Leipzig, 1734), 1:163; transcription in T 1716. Leibniz's corrections to "Uranias" are largely lost: LH 39, 18, Bl. 39–53, of which Bl. 41–53 are lost (per Editionskatalog / Ritterkatalog: missing since 1945). Four years later the opulent poetic work appears in print: Johann Wilhelm Petersen, *Uranias qua opera Dei magna* ... (Frankfurt/Leipzig, 1720). Samples show that Leibniz's corrections were incorporated, albeit without attribution, meaning Leibniz could be described as a secret co-author, at least for certain sections.

35   The discussion begins in Leibniz to Adolph-Theobald Overbeck, [June 17, 1715], LBr. 705 Bl. 39r.

36   Leibniz, "Geschichte als Wiederkehr des Gleichen? Das Apokatastasis-Fragment" (1715), in *Gottfried Wilhelm Leibniz: Schrifte und Briefe zur Geschichte*, trans. Malte-Ludolf Babin and ed. Gerd van den Heuvel (Hanover: Hahnsche, 2004), 550–61; Latin edition with further notes on textual and publication history.

37   Leibniz, "Geschichte als Wiederkehr des Gleichen?," 556.

38   For an interpretation of this line of thinking, cf. Hans Blumenberg, "Eine imaginäre Universalbibliothek," *Akzente* 28 (1981): 27–40; and Michel Fichant, "Ewige Wiederkehr oder unendlicher Fortschritt: Die Apokatastasisfrage bei Leibniz," *Studia Leibnitiana* 23, no. 2 (1991): 133–50.

39   Leibniz, "Geschichte als Wiederkehr des Gleichen?," 560–61.

40   Louis Bourguet to Leibniz, August 17, 1716, Niedersächsische Staats- und Universitätsbibliothek Göttingen, 2 Cod. Ms. Philos. 138 Bl. 17–18; transcription in T 1716.

41   Leibniz to Theobald Schöttel, July 2, 1716. For more on these matters, cf. Margot Faak, *Leibniz als Reichshofrat* (Berlin: Springer Spektrum, 2016), 101–3.

Epilogue

1   Paul Ritter, "Bericht eines Augenzeugen über Leibnizens Tod und Begräbnis," *Zeitschrift des Historischen Vereins für Niedersachsen* 81 (1916): 248. The last piece of writing of Leibniz's dated by him is a memorandum concerning the founding of a "Societas Subscriptionum" for the regulation of the book market in Germany, October 28, 1716, LH 39 Bl. 1–2.

2   Ritter, "Bericht eines Augenzeugen," 249.

3   Johann Bernoulli to Leibniz, December 5, 1716, LK-MOW Bernoulli 20 (previously: LBr. 57,2) VBl. A249–A250; transcription in T 1716.

4   Gottfried Teuber to Leibniz, November 3, 1716, LBr. 916 Bl. 38; transcription in T 1716. For more on the history of Leibniz's calculator, cf.

Gottfried Wilhelm Leibniz Bibliothek, ed., *Das letzte Original: Die Leibniz-Rechenmaschine der Gottfried Wilhelm Leibniz Bibliothek* (Hanover: Gottfried Wilhelm Leibniz Bibliothek, 2014); and Florin-Stefan Morar, "Reinventing Machines: The Transmission History of the Leibniz Calculator," *British Journal for the History of Science* 48, no. 1 (2015): 123–46.

5   Christian Breithaupt, *Zufällige Gedancken über die Methode, wie ein Atheist von der Existentz Gottes, und der Wahrheit der heiligen Schrifft zu überzeugen* . . . (Helmstedt, 1732), 56. My thanks to Siegmund Probst for pointing me to this passage.

BIBLIOGRAPHY

Aiton, Eric J. "An Unpublished Letter of Leibniz to Sloane." *Annals of Science* 38, no. 1 (1981): 103–7.

Anonymous. *Die heutigen Christlichen Souverainen von Europa, Das ist: Ein kurtzer Genealogischer und Politischer Abriß . . .* Breslau, 1704.

Antognazza, Maria Rosa. *Leibniz: An Intellectual Biography.* Cambridge: Cambridge University Press, 2009.

Armgardt, Matthias. *Das rechtslogische System der "Doctrina Conditionum" von Gottfried Wilhelm Leibniz.* Marburg: Elwert, 2001.

Arthur, Richard T. W. "Leibniz, Organic Matter and Astrobiology." In *Tercentenary Essays on the Philosophy and Science of Leibniz,* edited by Lloyd Strickland et al., 81–107. Cham: Palgrave Macmillan, 2017.

Babin, Malte-Ludolf. "Leibniz und der trimalcion moderne: Edition der Berichte von der Aufführung im Februar 1702." In *Studien zu Petron und seiner Rezeption / Studie su Petronio e sulla sua fortuna,* edited by Luigi Castagna and Eckhard Lefèvre, 331–60. Berlin: Walter de Gruyter, 2007.

Bayle, Pierre. "Pensées diverses I, §V." *Œuvres diverses, tome troisième, première partie.* The Hague, 1727: 11 B.

Beiderbeck, Friedrich, Irene Dingel, and Wenchao Li, eds. *Umwelt und Weltgestaltung: Leibniz' politisches Denken in seiner Zeit.* Göttingen: Vandenhoek und Ruprecht, 2014.

Beuys, Barbara. *Sophie Charlotte: Preußens erste Königen.* Berlin: Insel Verlag, 2018.

Biller, Gerhard. "Einleitung." A II, 3: xxii–xcvii.

Binswanger, Ludwig. *Wandlungen in der Auffassung und Deutung des Traumes: Von den Griechen bis zur Gegenwart.* Berlin: Springer, 1928.

Blumenberg, Hans. "Eine imaginäre Universalbibliothek." *Akzente* 28 (1981): 27–40.

Bodemann, Eduard, ed. *Aus den Briefen Herzogin Elisabeth Charlotte von Orléans an die Kurfürstin Sophie von Hanover: Ein Beitrag zur Kulturgeschichte des 17. und 18. Jahrhunderts, Hannover 1891.* Hanover: Hahn, 1891.

Böhmer, Georg Wilhelm. "Leibnizens Bild von ihm selbst entworfen." *Magazin für das Kirchenrecht, die Kirchen-und Gelehrten-Geschichte nebst Beiträgen zur Menschenkenntniß überhaupt* (1787), 1:104–7.

Bormastino, Antonio. *Historische Erzehlung von der Käyserlichen Residentz-Stadt Wienn* . . . Vienna, 1715.

Boyer, Christian. "Le plus petit cube magique." *La recherche: L'actualité des sciences* 373 (March 2004): 48–50.

Brancato, Mattia. "Leibniz, Weigel and the Birth of Binary Arithmetic." *Lexicon philosophicum: International Journal for the History of Texts and Ideas* 4 (2016): 151–72.

Breger, Herbert. "Leibniz' binäres Zahlensystem als Grundlage der Computertechnologie." *Jahrbuch der Akademie der Wissenschaften zu Göttingen* (2008): 385–91.

———. "Leibniz's Calculation with Compendia." In *Kontinuum, Analysis, Informales—Beiträge zur Mathematik und Philosophie von Leibniz*, edited by Wenchao Li, 147–58. Berlin: Springer, 2016.

———. "Die mathematisch-physikalische Schönheit bei Leibniz." In *Kontinuum, Analysis, Informales—Beiträge zur Mathematik und Philosophie von Leibniz*, edited by Wenchao Li, 105–13. Berlin: Springer, 2016.

Breithaupt, Christian. *Zufällige Gedancken über die Methode, wie ein Atheist von der Existentz Gottes, und der Wahrheit der heiligen Schrifft zu überzeugen* . . . Helmstedt: 1732.

Carhart, Michael C. *Leibniz Discovers Asia: Social Networking in the Republic of Letters*. Baltimore: John Hopkins University Press, 2019.

"Commercium epistolicum D. Johannis Collins, et aliorum de analysi promota: jussu societatis regiae in lucem editum." London: 1712 [1713].

Corbin, Alain. *The Foul and the Fragrant: Odor and the French Social Imagination*. Cambridge Mass.: Harvard University Press, 1986.

Crippa, Davide. *The Impossibility of Squaring the Circle in the 17th Century: A Debate Among Gregory, Huygens and Leibniz*. Cham: Birkhäuser, 2019.

Davis, Martin. *The Universal Computer: The Road from Leibniz to Turing*. New York: Norton, 2000.

De Risi, Vincenzo. *Geometry and Monadology: Leibniz's "Analysis Situs" and Philosophy of Space*. Basel: Birkhäuser, 2007.

———. *Leibniz on the Parallel Postulate and the Foundations of Geometry: The Unpublished Manuscripts*. Heidelberg: Springer, 2016.

Descartes, René. *Meditations de prima philosophia / Meditationen über die erste Philosophie*. Translated by Gerhart Schmidt. Stuttgart: Reclam, 1986.

Des Maizeaux, Pierre, ed. *Recueil de diverses pieces, sur la philosophie, la religion naturelle, l'histoire, les mathematiques, &c.* Amsterdam, 1720.

Doebner, Richard, ed. *Leibnizens Briefwechsel mit dem Minister von Bernstorff und andere Leibniz betreffende Briefe und Aktenstücke aus den Jahren 1705–1706.* Hanover: Hahn, 1882.

Duchesneau, François. *Organisme et corps organique de Leibniz à Kant.* Paris: Librairie Philosophique J. Vrin, 2018.

Eckert, Horst. *Gottfried Wilhelm Leibniz' Scriptores Rerum Burnsvicensium: Entstehung und historiographische Bedeutung.* Frankfurt: V. Klostermann, 1971.

Eckhart, Johann Georg von. "Lebensbeschreibung des Freyherrn von Leibnitz." In *Journal zur Kunstgeschichte und zur allgemeinen Literatur, 7. Theil*, edited by Christoph Gottlieb von Murr, 123–231. Nürnberg, 1779. Reprinted in J. A. Eberhard and J. G. Eckhart, eds. *Leibniz-Biographien.* Hildesheim: George Olms, 1982.

Ekirch, A. Roger. *At Day's Close: Night in Times Past.* New York: Norton, 2005.

Elster, Jon. *Leibniz et la formation de l'esprit capitaliste.* Paris: Aubier Montaigne, 1975.

Enzensberger, Hans Magnus. "G.W.L. (1646–1716)." In Enzensberger, *Mausoleum: Siebenunddreißig Balladen aus der Geschichte des Fortschritts.* Frankfurt: Suhrkamp, 1975.

Faak, Margot. *Leibniz als Reichshofrat.* Berlin: Springer Spektrum, 2016.

Fenves, Peter. *Arresting Language: From Leibniz to Benjamin.* Stanford: Stanford University Press, 2001.

Fichant, Michel. "Ewige Wiederkehr oder unendlicher Fortschritt: Die Apokatastasisfrage bei Leibniz." *Studia Leibnitiana* 23, no. 2 (1991): 133–50.

Fischer, Georg. "Der Philosoph Leibniz über Baracken." *Deutsche Zeitschrift für Chirurgie* 19 (1884): 135–36.

Franke, Konrad. "Zacharias Conrad von Uffenbach als Handschriftensammler. Ein Beitrag zur Kulturgeschichte des 18. Jahrhunderts." *Börsenblatt für den deutschen Buchhandel, Frankfurter Ausgabe* 51 (June 29, 1965): 1235–38.

Friedrich, Markus, and Monika Müller, eds. *Zacharias Konrad von Uffenbach: Büchersammler und Polyhistor in der Gelehrtenkultur um 1700.* Berlin: Walter de Gruyter, 2020.

Gädeke, Nora, "L'affaire de Monsieur Kortholt oder: Leibniz undercover— Eine Miszelle aus der Praxis der Leibnizedition." *Studia Leibnitiana* 41 (2011): 233–47.

———. "Leibniz' Korrespondenz im letzten Lebensjahr: Gerber Reconsidered." In *1716—Leibniz' letztes Lebensjahr: Unbekanntes zu einem bekannten Universalgelehrten*, edited by Michael Kempe, 83–109. Hanover: Gottfried Wilhelm Leibniz Bibliothek, 2016.

Gädeke, Nora, ed. *Leibniz als Sammler und Herausgeber historischer Quellen.* Wiesbaden: Harrasowitz Verlag, 2012.

Garber, Daniel. "Steno, Leibniz, and the History of the World." In *Steno and the Philosophers*, edited by Raphaële Andrault and Mogens Lærke, 201–29. Leiden: Brill, 2018.

Garrioch, David. "Sounds of the City: The Soundscape of Early Modern European Towns," *Urban History* 30 (May 2003): 5–25.

Gerland, Ernst. *Leibnizens nachgelassene Schriften physikalischen, mechanischen, und technischen Inhalts.* Leipzig: B.G. Teubner, 1906.

Glaser, Rüdiger. *Klimageschichte Mitteleuropas: 1200 Jahre Wetter, Klima, Katastrophen*, 2nd ed. Darmstadt: Wissenschaftliche Buchgemeinschaft, 2008.

Gödel, Kurt. "Über formal unentscheidbare Sätze der Principia Mathematica und verwandter Systeme I." *Monatshefte für Mathematik und Physik* 38 (1931): 173–98.

Göhrlich, Ekkehard. "Leibniz als Mensch und Kranker." Medical dissertation. Hanover, 1987.

Goldenbaum, Ursula. "Eine frühe Rezeption von Leibniz' Dynamik oder was wir von der Korrespondenz zwischen Louis Bourguet und Jacob Hermann lernen können." In *Leibniz in Latenz: Überlieferungsbildung als Rezeption (1716–1740)*, edited by Nora Gädeke and Wenchao Li, 223–53. Stuttgart: Franz Steiner Verlag, 2017.

Gössmann, Elisabeth. *"Die Päpstin Johanna": Der Skandal eines weiblichen Papstes: Eine Rezeptionsgeschichte*, 3rd ed. Munich: Aufbau Taschenbuch Verlag, 1998.

Gottfried Wilhelm Leibniz Bibliothek, ed. *Das letzte Original: Die Leibniz-Rechenmaschine der Gottfried Wilhelm Leibniz Bibliothek.* Hanover: Gottfried Wilhelm Leibniz Bibliothek, 2014.

Guerrier, Woldemar. *Leibniz in seinen Beziehungen zu Russland und Peter dem Grossen: Eine geschichtliche Darstellung dieses Verhältnisses nebst den darauf bezüglichen Briefen und Denkschriften.* St. Petersburg, 1873.

Hailperin, Theodore. "Algebraic Logic: Leibniz and Boole." In *A Boole Anthology: Recent and Classical Studies in the Logic of George Boole*, edited by James Gasser, 129–38. Dordrecht: Kluwer Academic, 2000.

Herse, Wilhelm. "Leibniz und die Päpstin Johanna." In *Beiträge zur Leibniz-Forschung*, edited by Georgi Schischkoff, 153–58. Reutlingen: Gryphius-Verlag, 1947.

Heuvel, Gerd van den. "Einleitung." In *Gottfried Wilhelm Leibniz: Schriften und Briefe zur Geschichte*, translated by Malte-Ludolf Babin and edited by Gerd van den Heuvel, 13–50. Hanover: Hansche, 2004.

———. "Die germanische Frühzeit als beste der möglichen Welten? Historiografie und Geschichtsauffassung bei Justus Möser und Gottfried Wilhelm Leibniz." In *Perspektiven der Landesgeschichte: Festschrift für Thomas Vogtherr*, edited by Christine van den Heuvel et al., 409–26. Göttingen: Wallstein Verlag, 2020.

———. "Leibniz im Netz: Die frühneuzeitliche Post als Kommunikationsmedium der Gelehrtenrepublik um 1700." *Lesesaal: Kleine Spezialitäten aus der Gottfried Wilhelm Leibniz Bibliothek—Niedersächsische Landesbibliothek* 32 (2009): 19–26.

———. "Leibniz und die Sulzbacher Protagonisten Christian Knorr von Rosenroth und Franciscus Mercurius van Helmont." *Morgen-Glantz: Zeitschrift der Christian Knorr von Rosenroth-Gesellschaft* 11 (2001): 77–104.

Hirsch, Eike Christian. *Der berühmte Herr Leibniz: Eine Biographie*. Munich: C.H. Beck, 2016.

Hofmann, Karl Gottlieb, ed. *Pantheon der Deutschen, 2. Teil*. Chemnitz, 1795.

Hooke, Robert. *Micrographia, or Some Physiological Descriptions of Minute Bodies Made by Magnifying Glasses . . .* London, 1667.

Hossenfelder, Sabine. *Das hässliche Universum: Warum unsere Suche nach Schönheit die Physik in die Sackgasse führt*. Frankfurt: Fischer, 2018.

Jost, Jürgen. *Leibniz und die moderne Naturwissenschaft*. Berlin: Springer, 2019.

Kabitz, Willy. *Leibniz und Berkeley*. Berlin: Sitzungsberichte der Preußischen Akademie der Wissenschaften, Phil. Hist. Klasse, 1932.

Kant, Immanuel. "Über das Mißlingen aller philosophischen Versuche in der Theodizee." In *Schriften zur Anthropologie, Geschichtsphilosophie, Politik und Pädagogik*, vol. 9, pt. 1 of *Werke*, edited by Wilhelm Weischedel, 105–24. Darmstadt: Wissenschaftliche Buchgesellschaft, 1983.

Kaplan, Robert. *The Nothing That Is: A Natural History of Zero*. Oxford: Oxford University Press, 1999.

Kauffmann, Kai. *"Es ist nur ein Wien!" Stadtbeschreibungen von Wien 1700 bis 1873. Geschichte eines literarischen Genres der Wiener Publizistik*. Vienna: Böhlau, 1994.

Keblusek, Marika, and Vera Noldus Badeloch, eds. *Double Agents: Cultural and Political Brokerage in Early Modern Europe*. Leiden: Brill, 2011.

Kempe, Michael. " 'Schon befand ich mich in Gedanken in Rußland . . . ': Johann Jakob Scheuchzer im Briefwechsel mit Gottfried Wilhelm Leibniz." In *Alte Löcher—neue Blicke. Zürich im 18. Jahrhundert: Aussen- und Innenperspektiven*, edited by Helmuth Holzhey and Simone Zurbuchen, 283–97. Zurich: Chronos-Verlag, 1997.

———. "Dr. Leibniz, oder wie ich lernte, die Bombe zu lieben. Zum Verhältnis von Wissenschaft und Militärtechnik in Europa um 1700." In *Der Philosoph im U-Boot: Praktische Wissenschaft und Technik im Kontext von Gottfried Wilhelm Leibniz*, edited by Michael Kempe, 113–45. Hanover: Gottfried Wilhelm Leibniz Bibliothek, 2015.

———. "Leibniz' Vorstellungen von europäischer und globaler Politik: Der ägyptische Plan revisited." In *Leibniz in Mainz: Europäische Dimensionen der Mainzer Wirkungsperiode*, edited by Irene Dingel et al., 73–91. Göttingen: Vandenhoeck & Ruprecht, 2019.

Kempe, Michael, ed. *1716—Leibniz' letztes Lebensjahr: Unbekanntes zu einem bekannten Universalgelehrten*. Hanover: Gottfried Wilhelm Leibniz Bibliothek, 2016.

Ker of Kersland, John. *The Memoirs of John Ker, of Kersland*. London, 1726.

Kliege-Biller, Herma. "Kirschsaft, Wien und Milchkaffee. Annotatiunculae culinariae." In *Individuation, Sýmpnoia pánta, Harmonia, Emanation: Festgabe für Heinrich Schepers zu seinem 75. Geburtstag*, edited by Klaus D. Dutz, 157–72. Münster: Nodus, 2000.

Kliege-Biller, Herma, Stephan Meier-Oeser, and Stephan Waldhoff. "Einen barocken Universalgelehrten edieren: Gottfried Wilhelm Leibniz, Sämtliche Schriften und Briefe." *Deutsche Zeitschrift für Philosophie* 64, no. 6 (2016): 951–97.

Knobloch, Eberhard. "Die entscheidende Abhandlung von Leibniz zur Theorie linearer Gleichungssysteme." *Studia Leibnitiana* 4 (1972): 163–80.

Kortholt, Christian. *Epistolae ad diversos . . .* Leipzig, 1734.

Krämer, Sybille. *Berechenbare Vernunft: Kalkül und Rationalismus im 17. Jahrhundert*. Berlin: Walter de Gruyter, 1991.

———. "Einführung: Wie aus 'nichts' etwas wird: Zur Kreativität der Null." In *Kreativität: XX. Deutscher Kongress für Philosophie 26.—30. September 2005 an der Technischen Universität Berlin*, edited by Günter Abel, 341–43. Hamburg: Meiner, 2006.

———. "Leibniz ein Vordenker der Idee des Netzes und des Netzwerks?" In *Vision als Aufgabe: Das Leibniz-Universum im 21. Jahrhundert*, edited by Martin Grötschel et al., 47–59. Berlin: Berlin-Brandenburgische Akademie der Wissenschaft, 2016.

Lamarra, Antonio, Roberto Palaia, and Pietro Pimpinella. *Le Prime traduzioni della Monadologie di Leibniz (1720–1721): Introduzione storico-critica, sinossi dei testi, concordanze contrastive*. Florence: Leo S. Olschki, 2001.

Lazardzig, Jan. *Theatermaschine und Festungsbau: Paradoxien der Wissensproduktion im 17. Jahrhundert*. Berlin: Akademie Verlag, 2007.

Lea, Kathleen M. "Sir Anthony Standen and Some Anglo-Italian Letters." *English History Review* 47 (1932): 466–77.

Leibniz, Gottfried Wilhelm. "Additio ad schedam in Actis proxime antecedentis Maji pag. 233 editam, de dimensionibus curvilineorum, per G.G.L." *Acta eruditorum* (December 1684): 585–87. German translation: Leibniz, "Ergänzung zum Beitrag *Wie man zur Ausmessung von Figuren kommt*, der in den *Acta* des letztvergangenen Mai auf S. 233 f. veröffentlicht ist," *Die mathematischen Zeitschriftenartikel*. Translated by Heinz-Jürgen Heß and Malte-Ludolf Babin, 47–48. Hildesheim: Georg Olms, 2011.

———. "Anfangsgründe einer allgemeinen Charakteristik" (1677). In *Schriften zur Logik und zur philosophischen Grundlegung von Mathematik und*

Naturwissenschaft, vol. 4 of *Philosophische Schriften*, translated and edited by Herbert Herring. Frankfurt: Insel Verlag, 1992.

———. "Brevis demonstratio erroris memorabilis Cartesii . . . ." *Acta eruditorum* (March 1686): 161–63.

———. *Der Briefwechsel mit Bartholomäus Des Bosses*. Translated and edited by Cornelius Zeheter. Hamburg: F. Meiner, 2007.

———. *Der Briefwechsel mit den Jesuiten in China (1689–1714)*, edited by Rita Widmaier. Text prepared and translated by Malte-Ludolf Babin. Hamburg: Meiner, 2006.

———. *Essais de Théodicée sur la bonté de dieu, la liberté, et de l'homme et l'origine du mal*. Amsterdam, 1710.

———. "Extrait du dictionnaire de M. Bayle article Rorarius p. 2599 sqq. de l'edition de l'an 1702 avec mes remarques." GP 4:524–54.

———. "Flores sparsi in tumulum papissae." In *Bibliotheca Historica Goettingensis*, edited by Christian Ludwig Scheidt, 1:297–392. Göttingen/Hanover, 1758.

———. "Geschichte als Wiederkehr des Gleichen? Das Apokatastasis-Fragment (1715)." In *Gottfried Wilhelm Leibniz: Schriften und Briefe zur Geschichte*, translated by Malte-Ludolf Babin and edited by Gerd van den Heuvel, 550–61. Hanover: Hahnsche, 2004.

———. *Gottfried Wilhelm Leibniz: Schriften und Briefe zur Geschichte*, translated by Malte-Ludolf Babin and edited by Gerd van den Heuvel. Hanover: Hahnsche, 2004.

———. "Günstiger Leser." Foreword to Anicius Manlius Severinus Boethius, *Consolatio Philosophiae oder Christlich-vernufft-gemesser Trost und Unterricht in Widerwertigkeit und Bestürzung über dem vermeinten Wohl- oder Übel-Stand der Bösen und Frommen . . .* Translated by Franciscus Mercurius Helmont. Lüneburg, 1697.

———. "Imago Leibnitii." In *Leben und Werk von Gottfried Wilhelm Leibniz: Eine Chronik*, edited by Kurt Müller and Gisela Krönert, 1–2. Frankfurt: Klostermann, 1969.

———. "In der Vernunft begründete Prinzipien der Natur und der Gnade." In *Kleine Schriften zur Metaphysik*, vol. 1 of *Philosophische Schriften*, translated and edited by Hans Heinz Holz, 412–39. Frankfurt: Surhkamp, 1996.

———. *Leibniz's Monadology*. Translated by Lloyd Strickland. Edinburgh: Edinburgh University Press, 2014.

———. "Metaphysische Abhandlung" (1686). In *Kleine Schriften zur Metaphysik*, vol. 1 of *Philosophische Schriften*. Translated and edited by Hans Heinz Holz, 56–172. Frankfurt: Suhrkamp, 1996.

———. "Principes de la nature et de la grâce fondés en raison." GP 6:598–606.

———. "Principles of Nature and of Grace, Founded on Reason," in Leibniz,

Philosophical Writings, edited by G.H.R. Parkinson. Translated by Mary Morris and G.H.R. Parkinson, 195–204. London: Dent, 1973.

———. *Protogaea. Sive de prima facie telluris et antiquissimae historiae vestigiis in ipsis naturae monumentis dissertatio . . .* Göttingen, 1749.

———. "Protogaea autore G.G.L." *Acta eruditorum* (January 1693): 40–42.

———. "[Selbstschilderung. G.G.L.]." In *Leibnizens geschichtliche Aufsätze und Gedichte aus den Handschriften der Königlichen Bibliothek zu Hannover*, edited by Georg Heinrich Pertz, 4:173–75. Hanover: Im Verlage der Hahnschen Hofbuchhandlung,1847.

———. "Système nouveau de la nature et de la communication des substances, aussi vien que de l'union qu'il y a entre l'âme et le corps." *Journal des Sçavans* (June 27, 1695): 294–300, and (July 4, 1695): 301–6.

———. "Tagebuch [August 1696–April 1697]." In *Leibnizens geschichtliche Aufsätze und Gedichte aus den Handschriften der Königlichen Bibliothek zu Hannover*, edited by Georg Heinrich Pertz. Hanover: Im Verlage der Hahnschen Hofbuchhandlung, 1847.

———. *Theodicy, Abridged.* Translated by E. M. Huggard. Ontario: J.M. Dent & Sons, 1966.

———. *Die Theodizee von der Güte Gottes, der Freiheit der Menschen und dem Ursprung des Übels*, vols. 2/1 and 2/2 of *Philosophische Schriften*. Translated and edited by Herbert Herring. Frankfurt: Suhrkamp, 1996.

———. "Über die Kontingenz." In *Kleine Schriften zur Metaphysik*, translated and edited by Hans Heinz Holz, 179–87. Frankfurt: Suhrkamp, 1996.

———. "Von dem Verhängnisse" (1695). In *Hauptschriften zur Grundlegung der Philosophie*, ed. Ernst Cassirer and Artur Buchenau, 2:337–42. Hamburg: F. Meiner, 1996.

Leibniz, Gottfried Wilhelm, and Antoine Arnauld. *Der Briefwechsel mit Antoine Arnauld*, edited by Reinhard Finster. Hamburg: F. Meiner, 1997. English translation: Leibniz and Arnauld. *The Leibniz-Arnauld Correspondence*, translated and edited by H. T. Mason. Manchester: Manchester University Press, 1967.

Leibniz, Gottfried Wilhelm, and Electress Sophie of Hanover. *Briefwechsel*, edited by Wenchao Li. Translated by Gerda Utermöhlen and Sabine Sellschopp. Göttingen: Wallstein Verlag, 2017.

Mackensen, Ludolf von. "Die ersten dekadischen und dualen Rechenmaschinen." In *Gottfried Wilhelm Leibniz. Das Wirken des großen Universalgelehrten als Philosoph, Mathematiker, Physiker, Techniker*, edited by Karl Popp and Edwin Stein, 85–100. Hanover: Schlütersche, 2000.

Mayer, Uwe. "Mündliche Kommunikation und schriftliche Überlieferung—Die Gesprächsaufzeichnungen von Leibniz und Tschirnhaus zur Infinitesimalrechnung aus der Pariser Zeit." In *VIII. Internationaler Leibniz Kongress:*

Einheit in der Vielfalt. Vorträge 1. Teil, edited by Herbert Breger, Jürgen Herbst, und Sven Erdner, 588–94. Hanover, 2006.

Meckseper, Cord. *Das Leibnizhaus in Hannover: Die Geschichte eines Denkmals.* Hanover: Schlütersche, 1983.

Meder, Stephan. *Der unbekannte Leibniz: Die Entdeckung von Recht und Politik durch Philosophie.* Vienna: Böhlau Verlag, 2018.

Meier, Monika. "Leibniz' Briefwechsel mit Caroline von Ansbach, Princess of Wales—'Leibniz' in Großbritannien und neue ökumenische Initiativen." In *1716—Leibniz' letztes Lebensjahr: Unbekanntes zu einem bekannten Universalgelehrten*, edited by Michael Kempe, 241–91. Hanover: Gottfried Wilhelm Leibniz Bibliothek, 2016.

Meier-Oeser, Stephan. " 'Adieu le Vuide, les Atomes, et toute la Philosophie de M. Newton.' Zur Leibniz-Clarke-Kontroverse." In *1716—Leibniz' letztes Lebensjahr: Unbekanntes zu einem bekannten Universalgelehrten*, edited by Michael Kempe, 293–317. Hanover: Gottfried Wilhelm Leibniz Bibliothek, 2016.

Morar, Florin-Stefan. "Reinventing Machines: The Transmission History of the Leibniz Calculator." *British Journal for the History of Science* 48, no. 1 (2015): 123–46.

Müller, Kurt, and Gisela Krönert, eds. *Leben und Werk von Gottfried Wilhelm Leibniz: Eine Chronik.* Frankfurt: Klostermann, 1969.

Münzenmayer, Hans Peter. "Leibniz' Inventum Memorabile: Die Konzeption einer Drehzahlregelung vom März 1686." *Studia Leibnitiana* 8 (1976): 113–19.

Murr, Christoph Gottlieb von. "Einige Zusätze zum Eckhartischen Lebenslaufe des Herrn von Leibnitz" (1779). In *Leibniz-Biografien*, edited by J. A. Eberhard and J. G. Eckhart, 204–31. Hildesheim: George Olms, 1982.

Nachtomy, Ohad. "Infinity and Life: The Role of Infinity in Leibniz's Theory of Living Beings." In *The Life Sciences in Early Modern Philosophy*, edited by Ohad Nachtomy and Justin E. H. Smith, 9–28. Oxford: Oxford University Press, 2014.

———. "A Tale of Two Thinkers, One Meeting, and Three Degrees of Infinity: Leibniz and Spinoza (1675–78)." *British Journal for the History of Philosophy* 19 (2011): 935–61.

Neumann, Hanns-Peter. "Monadentheorie und Monadologie." In *Gottfried Wilhelm Leibniz: Rezeption, Forschung, Ausblick*, edited by Friedrich Beiderbeck, Wenchao Li, and Stephan Waldhoff, 500–7. Stuttgart: Franz Steiner Verlag, 2020.

O'Hara, James G. " 'J'aime mieux un Leewenhoek qui me dit ce qu'il voit, qu'un Cartesien qui me dit ce qu'il pense,' Leibniz, Leeuwenhoek und die Entwicklung der Experimentellen Naturwissenschaft." In *1716—Leibniz' letztes Lebensjahr: Unbekanntes zu einem bekannten Universalgelehrten*, edited by Michael Kempe, 145–75. Hanover: Gottfried Wilhelm Leibniz Bibliothek, 2016.

———. *Leibniz's Correspondence in Science, Technology and Medicine (1676–1701)* (Leiden: Brill, 2024).

Ottaviani, Osvaldo, and Alessandro Becchi, eds. "Leibniz on Animal Generation: An Unpublished Text (LH 37, 7 Bl. 6–7) with Introduction, Critical Edition, Translation, and Commentary," *Leibniz Review* 30 (2020): 63–106.

Pape, Ingetrud. "Von den 'möglichen Welten' zur 'Welt des Möglichen'. Leibniz im modernen Verständnis." In *Akten des Internationalen Leibniz-Kongresses, Hannover, 14.–19. November 1966*, vol. 1, *Metaphysik—Monadenlehre*, edited by Kurt Müller and Wilhelm Totok, 266–87. Wiesbaden: F. Steiner, 1968.

Peckhaus, Volker. *Logik, Mathesis universalis und allgemeine Wissenschaft: Leibniz und die Wiederentdeckung der formalen Logik im 19. Jahrhundert.* Berlin: Akademie Verlag, 1997.

Peres, Constanze. "Neuheit und Kreativität—analogisches Denken und Leibniz' Idee der Erfindung. Zur Einführung." In *Wie entsteht Neues? Analogisches Denken, Kreativät und Leibniz' Idee der Erfindung*, edited by Constanze Peres. Paderborn: Wilhelm Fink Verlag, 2020: 7–12.

Perler, Dominik. "Was ist eine Person? Überlegungen zu Leibniz." *Deutsche Zeitschrift für Philosophie* 64, no. 3 (2016): 329–51.

Petersen, Johann Wilhelm. *Uranias qua opera Dei magna . . .* Frankfurt/Leipzig, 1720.

Pfister, Christian, and Walter Bareiss. "The Climate in Paris between 1675 and 1715 According to the Meteorological Journal of Louis Morin." In *Climatic Trends and Anomalies in Europe 1675–1715: High Resolution Spatio-Temporal Reconstructions from Direct Meteorological Observations and Proxy Data: Methods and Results*, edited by Burkhard Frenzel et al., 151–72. Stuttgart: G. Fischer, 1994.

Piel, Gisela, and Susanne Luber. "Ein 'berühmtes Brunnenjahr.' Zar Peter I. in Pyrmont im Juni 1716." In *Zar Peter der Grosse: Die zweite große Reise nach Westeuropa, 1716–1717: Auf Europäischer Bühne*, edited by Dieter Alfter, 87–96. Hameln: Niemeyer, 1999.

Probst, Siegmund. "The Calculus." In *The Oxford Handbook of Leibniz*, edited by Maria Rosa Antognazza, 213–15. Oxford: Oxford Academic, 2018.

Pross, Addy. "Toward a General Theory of Evolution: Extending Darwinian Theory to Inanimate Matter." *Journal of Systems Chemistry* 2, no. 1 (2011) 1–14.

Rabouin, David. "A Fresh Look at Leibniz' Mathesis Universalis." In *"Für unser Glück oder das Glück anderer": Vorträge des X. Internationalen Leibniz-Kongresses, Hannover, 18.–23. Juli 2016*, edited by Wenchao Li and Ute Beckmann, 4:505–19. Hildesheim: Georg Olms Verlag, 2016.

Ramelow, Tilman. *Gott, Freiheit, Weltenwahl: Der Ursprung des Begriffes der besten aller möglichen Welten in der Metaphysik der Willensfreiheit zwischen Antonio Perez S. J. (1599–1649) und G. W. Leibniz (1646–1716).* Leiden: E.J. Brill, 1997.

Redi, Francesco. *Esperienze intorno alla generazione degli insetti*. 1668; reprinted Florence, 1996.

Reimmann, Jakob Friedrich. *Jacob Friederich Reimmanns, Weyland Hochverdienten Superintendentens der Evangelischen Kirchen . . . Eigene Lebens-Beschreibung Oder Historische Nachricht von Sich Selbst . . .* Brunswick, 1745.

Riley, Patrick. *Leibniz' "Monadologie" als Theorie der Gerechtigkeit.* Hanover: Wehrhahn Verlag, 2017.

Ritter, Paul. "Bericht eines Augenzeugen über Leibnizens Tod und Begräbnis." *Zeitschrift des Historischen Vereins für Niedersachsen* 81 (1916): 251.

Roll, Christine. "Barbaren? 'Tabula rasa'? Wie Leibniz sein neues Wissen über Russland auf den Begriff brachte; eine Studie über die Bedeutung und Vernetzung gelehrter Korrespondenzen für die Ermöglichung aufgeklärter Diskurse." In *Umwelt und Weltgestaltung: Leibniz' politisches Denken in seiner Zeit*, edited by Friedrich Beiderbeck et al., 307–58. Göttingen: Vandenhoek & Ruprecht, 2014.

Scheel, Günter. "Leibniz als politischer Ratgeber des Welfenhauses." In *Leibniz und Niedersachsen: Tagung anlässlich des 350. Geburtstages von G. W. Leibniz, Wolfenbüttel, 1996*, edited by Herbert Breger and Friedrich Niewöhner, 35–52. Stuttgart: Franz Steiner Verlag, 1999.

Schmidhuber, Jürgen. "Der erste Informatiker. Wie Gottfried Wilhelm Leibniz den Computer erdachte." *Frankfurter Allgemeine Zeitung*, no. 112 (May 17, 2021): 18.

Schmölzer, Heide. *Die Pest in Wien*. Innsbruck: Haymon Verlag, 2015.

Schüller, Volkmer, ed. *Der Leibniz-Clarke Briefwechsel*. Berlin: Akademie Verlag, 1991.

Seife, Charles. *Zero: The Biography of a Dangerous Idea*. New York: Viking, 2000.

Sellschopp, Sabine. " 'Eine kleine tour nach Hamburg incognito': Zu Leibniz' Bemühungen von 1701 um die Position eines Reichshofrats." *Studia Leibnitiana* 37, no. 1 (2005): 68–82.

Siegert, Bernhard. *Passagen des Digitalen: Zeichenpraktiken der neuzeitlichen Wissenschaft 1500–1900*. Berlin: Brinkmann & Bose, 2003.

Silvestri, Federico. "Due inediti dal carteggio Leibniz-Bourguet," *Rivista di storia della filosofia, nuova seria 3* (2018): 481–505.

Sloterdijk, Peter. *Philosophische Temperamente: Von Platon bis Foucault*. Munich: Diederichs, 2009.

Smith, Justin E. H. *Divine Machines: Leibniz and the Sciences of Life* (Princeton, NJ: Princeton University Press, 2011).

———. "Lebenswissenschaften." In *Gottfried Wilhelm Leibniz: Rezeption, Forschung, Ausblick*, edited by Friedrich Beiderbeck, Wenchao Li, and Stephan Waldhoff, 763–76. Stuttgart: Franz Steiner Verlag, 2020.

Steiner, Benjamin. "Leibniz, Colbert und Afrika. Wissen und Nicht-Wissen über die geopolitische Bedeutung des Kontinents um 1670." In *Wissenskulturen in der Leibniz-Zeit. Konzepte—Praktiken—Vermittlung*, edited by Friedrich Beiderbeck and Claire Gantet, 75–112. Berlin: De Gruyter Oldenbourg, 2021.

Stencil, Eric. "Arnauld's God Reconsidered." *History of Philosophy Quarterly* 36 (2019): 19–38.

Stock, Jürgen, and Rainer Weichert. *Die Hartzings: Der Aufstieg einer Moerser Familie unter der Ostindischen Kompanie*. Duisburg: Mercator Verlag, 2020.

Strack, Clara. *Ursprung und sachliches Verhältnis von Leibnizens sogenannter Monadologie und den Principes de la nature et de la grâce. 1. Teil: Die Entstehungsgeschichte der beiden Abhandlungen*. Berlin, 1915.

Strickland, Lloyd, ed. *Leibniz and the Two Sophies: The Philosophical Correspondence*. Toronto: Iter, 2011.

Strömmer, Elisabeth. *Klima-Geschichte: Methoden der Rekonstruktion und historische Perspektive: Ostösterreich 1700 bis 1830*. Vienna: Deuticke, 2003.

Stuber, Regina. "Leibniz' Bemühungen um Russland: Eine Annäherung." In *1716—Leibniz' letztes Lebensjahr*, edited by Michael Kempe, 203–39. Hanover: Gottfried Wilhelm Leibniz Bibliothek, 2016.

Trunk, Achim. "Sechs Systeme. Leibniz und seine *signa ambigua*." In *"Für unser Glück oder das Glück anderer." Vorträge des X. Internationalen Leibniz-Kongresses, Hannover, 18.–23. Juli 2016*, edited by Wenchao Li and Ute Beckmann, 4:191–207. Hildesheim: Georg Olms Verlag, 2016.

Uffenbach, Zacharias Konrad von. *Merkwürdige Reisen durch Niedersachsen, Holland und England*. Ulm/Memmingen, 1753.

Uhlig, Ingo. *Traum und Poiesis: Produktive Schlafzustände 1641–1810*. Göttingen: Wallstein Verlag, 2015.

Utermöhlen, Gerda. "Leibnizens Hundebittschrift, der Papinsche Topf und die Geschichte eines hartnäckigen Lesefehlers." *Hannoversche Geschichtsblätter, Neue Folge* 33 (1979): 57–62.

———. "Leibniz im kulturellen Rahmen des hanoverschen Hofes." In *Leibniz und Niedersachsen: Tagung anlässlich des 350. Geburtstages von G. W. Leibniz, Wolfenbüttel, 1996*, edited by Herbert Breger and Friedrich Niewöhner, 213–26. Stuttgart: Franz Steiner Verlag, 1999.

Valla, Lorenzo. *Über den freien Willen. De libero arbitrio*, edited by Eckhard Keßler. Munich: W. Fink, 1987.

Vogl, Joseph. "Leibniz, Kameralist." *Europa: Kultur de Sekretäre*, edited by Bernhard Siegert and Joseph Vogl. Zurich: Diaphanes Verlag, 2003: 97–109.

Voltaire. *Candide*. Translated by John Butt. Harmondsworth, Middlesex: Penguin Books, 1947.

Wahl, Charlotte. "Assessing Mathematical Progress: Contemporary Views on the Merits of Leibniz's Infinitesimal Calculus." In *Natur und Subjekt. Vorträge des*

IX. Internationalen Leibniz-Kongresses, 3. Teil, edited by Herbert Breger et al., 1172–82. Hanover: Gottfried-Wilhelm-Leibniz-Gesellschaft, 2011.

Wellmer, Friedrich-Wilhelm, and Jürgen Gottschalk. "Bergbau und Geologie." In *Gottfried Wilhelm Leibniz: Rezeption, Forschung, Ausblick,* edited by Friedrich Beiderbeck, Wenchao Li, and Stephan Waldhoff, 777–87. Stuttgart: Franz Steiner Verlag, 2020.

———. "Leibniz' Scheitern im Oberharzer Silberbergbau: Neu betrachtet, insbesondere unter klimatischen Gesichtspunkten." *Studia Leibnitiana* 42 (2010): 186–207.

Widmaier, Rita. "Die Dyadik in Leibniz' letztem Brief an Nicolas Remond." *Studia Leibnitiana* 49, no. 2 (2017): 139–76.

Wienerisches Diarium. Nr. 1155 (August 25–28, 1714).

Witsen, Nicolaas. *Nieuwe Lantkaarte van het noorder en ooster dell van Asa en Europa, strekkende van Nova Zemla tot China.* Amsterdam: 1687. GWLB: Ktr 118.

Wittgenstein, Ludwig. *Philosophische Untersuchungen,* vol. 2 of *Werkausgabe.* Frankfurt: Suhrkamp, 1984.

Zuse, Horst. "Der lange Weg zum Computer: Von Leibniz' Dyadik zu Zuses Z3." In *Vision als Aufgabe: Das Leibniz-Universum im 21. Jahrhundert,* edited by Martin Grötschel et al., 111–24. Berlin: Berlin-Brandenburgische Akademie der Wissenschaft, 2016.

SUGGESTED READING

(B = see Bibliography)

LOOKING FOR A WAY INTO THE LEIBNIZ UNIVERSE? GIVING reading tips is both easy and difficult at the same time. Easy, because the best and most direct point of entry is available online through the Academy Edition of Leibniz's works: https://leibnizedition.de. The volumes of the historical-critical edition, at least those that have been digitized or scanned, can be accessed directly from this website, free of charge; otherwise print editions are available from De Gruyter Verlag. It's difficult, however, because the Leibniz texts in the Academy Edition are only available in the original (mostly in Latin or French). The very informative introductions to each volume provide an excellent overview. Taken together, all the introductions of the different volumes of each series can be rightly described as the best German-language monograph available on the life and work of Leibniz, at least for the years covered by the Academy Edition. (So far the volumes have not yet reached the end of Leibniz's life.)

German translations of Leibniz's work are for the most part limited to selections of philosophically relevant letters and writings. The four-volume edition from Suhrkamp Taschenbuch, for example, edited and translated by Hans Heinz Holz and Herbert Herring, is readily available, as is the two-volume edition from Meiner, edited by Ernst Cassirer, translated by Arthur Buchenau. For Leibniz's essays on mathematics, the German translation from Heinz-Jürgen Heß and Malte-Ludolf Babin (B) is recommended; likewise, for selected

writings on history, the German edition from Malte-Ludolf Babin
(B). Leibniz's harshest critic, to the extent that the criticism applies
to Leibniz and not those who popularized him, Voltaire's *Candide*,
is available in German in a slim Reclam edition (B). Anyone who
wants a quick and brisk introduction to Leibniz is in good hands with
Reinhard Finster and Gerd van den Heuvel's introduction, put out by
Rowohlt (rororo). The biographies by Eike Christian Hirsch (B) and
Maria Rosa Antognazza (B) offer lengthier but still readable presentations of the life and work.

To those who wish to delve even deeper into the sheer boundless
cosmos of scholarly literature on Leibniz, first of all: be warned. Second of all, you'll have to be your own guide. Third—in my own sub-
jective view—you might begin here with Aron Gurwitsch, *Leibniz.
Philosophie des Panlogismus* (Berlin: Walter de Gruyter, 1974), a classic, or Gilles Deleuze, *Die Falte. Leibniz und der Barock* (Suhrkamp
Taschenbuch, 2023), another classic, contested, but still refreshingly
provocative. Further, there are Sybille Krämer's sharp-witted and
analytically in-depth *Berechenbare Vernunft* (B) and, with an inter-
esting art-historical perspective, Horst Bredekamp's *Die Fenster der
Monade: Gottfried Wilhelm Leibniz' Theater der Natur und Kunst*, rev.
3rd ed. (Berlin: Walter de Gruyter, 2020). For more on Leibniz's
"two Sophies," the biography of Queen Sophie Charlotte by Barbara
Beuys (B) is recommended, and as for Electress Sophie, the correspondence with Leibniz is available in German, edited by Wenchao
Li and translated by Gerda Utermöhlen and Sabine Sellschopp (B).
The letters between the electress and the scholar open up the cultural
world of the Baroque court—unvarnished (though heavily made-up)
and many-faceted—in all its dimensions, between politics and morality, mathematics and metaphysics, chitchat and gossip.

ILLUSTRATION CREDITS

13 Photo: GWLB.

20 GWLB, N-A 7051. This copy of the book comes from the library of Martin Fogel and was available to Leibniz in Hanover.

23 Graphic by Peter Palm, Berlin, from a drawing by Michael Kempe.

24 LH 35 VIII 18 Bl. 2v (excerpt).

40 LH 38 Bl. 313.

53 Graphic by Peter Palm, Berlin, from a drawing by Jürgen Gottschalk.

62 LH 41, 4 Bl. 1r.

69 GWLB, Mappe 18, XIX.C, Nr. 178b.

77 GWLB, N-A 7051 (for the provenance of this book see the credit for the image on p. 20, above).

80 GWLB, Gd-A 1246 1795v2.

82 GWLB, P-A 176 (*Marginalexemplar* Leibniz).

97 LH 38 Bl. 266.

101 GWLB, Leibniz-Archiv.

115 Enclosed with N. 318, available in A I, 20 N. 319, 555–56, LK-MOW Bouvet 10 Bl. 27–28.

126 GWLB, Leibniz-Archiv.

129 GWLB, Portrait of Leibniz, c. 1711 (after La Bonté, 1788; copy).

166 LBr. 768, Bl. 22 (excerpt).

170 GWLB, Leibn. 131.

184 GWLB, GA 4009.

197 GWLB, Leibn. 211.

207 LK-MOW Hermann 10 Bl. 117 (previously: LBr. 396 Bl. 117), here Bl. 117r.

INDEX

Page numbers in *italics* indicate illustrations.